530.1 ROUN

YALE COLLEGE
LEARNING RESOURCE CENTRE

D1148223

maths
for advanced physics

Coleg Ial Wrecsam / Yale College of Wrexham
Canolfan Adnoddau Dysgu / Learning Resource Centre
Ffon/Telephone: (01978) 311794 ext 2257
E ... il ...c@yale-wrexham.ac ...

... e last date sta... w

John Rounce

COLEG IAL WRECSAM
YALE COLLEGE OF WREXHAM
41670

Text © John Rounce 2002
Original illustrations © Nelson Thornes Ltd 2002

The right of John Rounce to be identified as authors of this work has been asserted by him in accordance with the Copyright, Designs and Patents Act 1988.

All rights reserved. No part of this publication may be reproduced or transmitted in any form or by any means, electronic or mechanical, including photocopy, recording or any information storage and retrieval system, without permission in writing from the publisher or under licence from the Copyright Licensing Agency Limited, of 90 Tottenham Court Road, London W1T 4LP.

Any person who commits any unauthorised act in relation to this publication may be liable to criminal prosecution and civil claims for damages.

Published in 2002 by:
Nelson Thornes Ltd
Delta Place
27 Bath Road
CHELTENHAM
GL53 7TH
United Kingdom

02 03 04 05 06 / 10 9 8 7 6 5 4 3 2 1

A catalogue record for this book is available from the British Library

ISBN 0 7487 6508 5

Illustrations by Oxford Designers and Illustrators
Page make-up by Mathematical Composition Setters Ltd

Printed and bound in Croatia by Zrinski

Contents

Introduction   vi

1 Multiplication and division   **1**
  1.1 Multiplying   1
  1.2 Division   6
  Exam questions   12
  Answers to Test Yourself Questions   14

2 Fractions   **15**
  2.1 Fractions and their names   15
  2.2 Simplifying expressions   17
  2.3 Combining fractions   21
  2.4 Decimal fractions   28
  2.5 Percentage   30
  Exam questions   32
  Answers to Test Yourself Questions   32

3 Brackets and negative numbers   **34**
  3.1 Brackets   34
  3.2 Plus, minus and directed numbers   37
  3.3 Some special uses of + and − signs and brackets in physics   41
  Exam questions   42
  Answers to Test Yourself Questions   43

4 Powers   **44**
  4.1 What powers will I meet?   44
  4.2 Using a calculator for powers of numbers other than 10   53
  4.3 The exponential function   54
  4.4 Standard form   56
  Exam questions   57
  Answers to Test Yourself Questions   58

5 Units   **60**
  5.1 The concept of units   60
  5.2 Dimensions – what are they?   67
  Exam questions   72
  Answers to Test Yourself Questions   73

6 Roots of numbers   **74**
  6.1 Roots   74
  6.2 Squaring to remove roots   77
  6.3 Powers for roots   78
  6.4 Roots $\sqrt{a \times b}$ and $\sqrt{\dfrac{a}{b}}$ 79
  6.5 Positive and negative roots   80

CONTENTS

Exam questions   82
Answers to Test Yourself Questions   82

7 Errors   **83**
7.1 Experimental errors   83
7.2 Significant figures   86
7.3 Combining errors   87
7.4 Allowing for errors when calculating   90
Exam questions   92
Answers to Test Yourself Questions   93

8 Algebra, equations and transpositions   **94**
8.1 Algebra – its meaning   94
8.2 Equations, formulae and identities   95
8.3 Proportionality   99
Exam questions   107
Answers to Test Yourself Questions   108

9 Graphs   **109**
9.1 Graphs – how to draw them   109
9.2 Equations for graphs   119
9.3 Graphs that are curves   122
9.4 Other uses for graphs   124
9.5 Some special graphs   130
Exam questions   133
Answers to Test Yourself Questions   134

10 Angles and geometry   **136**
10.1 Angles   136
10.2 Isosceles and equilateral, congruent and similar triangles   140
10.3 Pythagoras' formula   141
10.4 Circles and discs   144
10.5 Three-dimensional shapes   146
10.6 Angles measured in radians   149
Exam questions   151
Answers to Test Yourself Questions   152

11 Trigonometry   **153**
11.1 Trigonometric ratios – what are they?   153
11.2 Vectors   158
11.3 Resolving forces   162
11.4 Equilibrium   168
Exam questions   169
Answers to Test Yourself Questions   171

12 Averages   **172**
12.1 Mean values   172
12.2 Continuous variables   174
12.3 Root mean square values   177
Exam questions   181
Answers to Test Yourself Questions   181

13 Simultaneous equations and quadratic equations **182**
13.1 Types of equation 182
13.2 Simultaneous equations 183
13.3 Solving quadratic equations 188
Exam questions 192
Answers to Test Yourself Questions 193

14 Logarithms **194**
14.1 Logarithms and their properties 194
14.2 Rules concerning logarithms 198
Exam questions 204
Answers to Test Yourself Questions 205

15 Maths of circular motion, oscillations and waves **206**
15.1 Circular motion 206
15.2 Simple harmonic motions 208
15.3 Sines of angles greater than 90° 210
15.4 Sine curves and sine waves 212
15.5 Phasors 217
Exam questions 219
Answers to Test Yourself Questions 221

Appendices **222**
Appendix I Combining possible errors 222
Appendix II Proof of Pythagoras' theorem 224
Appendix III Tables of symbols, abbreviations and units 225

Answers to exam-type questions **226**
Index **228**
Acknowledgements **232**

Introduction

About this book

Whether you are studying Advanced Level Physics after getting an A grade in GCSE Mathematics or you feel much less certain about your understanding of maths having struggled at GCSE, this book will help you. It aims to give you more confidence in both the basic GCSE maths you need to get started in your physics studies and to provide the extra maths you will need for Advanced Level Physics.

Maths that is memorised but not understood is difficult to apply to the wide range of calculations met in physics. You can feel that maths is confusing, muddled and mysterious. The explanations, advice and experience obtainable from *Maths for Advanced Physics* should give you the competence and confidence that leads to an interesting, successful and enjoyable physics course.

Maths is important in physics. In exam type questions you will need it for calculations and when asked to prove a formula. Practical work also needs maths, and examples of how to present your calculations, draw graphs and display the units you use are included here.

How to use this book

You may decide to work steadily through all the text of this book or start with topics for which you need the most help. (You should note, however, that in any chapter a familiarity with earlier topics may be assumed.) The index at the back of the book will help you to locate any information you need from earlier chapters. Also the 'Contents' list at the front, together with the list of objectives at the start of each chapter, will guide you to the maths you want.

An alternative way of using the book is to try the exercises and through them discover what maths you still need to study.

You will find here all the maths topics required for Advanced Level Physics courses by the main awarding bodies of England and Wales. The requirements of similar physics courses should be covered as well. Each topic has a short chapter that has

- a list of objectives for you to achieve

- an introduction to put the topic into context

- an explanation

- some 'Test Yourself' questions with the answers to these at the end of the chapter

- some exam type questions at the end of each chapter with the answers to these at the back of the book

To help, we have used a simple code.

 Test Yourself These questions are in the text and you should try each one as you come to it. If you can do it, you have understood the point. If you cannot, it means that you need to think again about the text above the symbol.

KEY FACT *Shaded boxes enclose a key fact of a statement of principle which you should be able to remember and use.*

 shows a section on using a VPAM (visually perfect algebraic method) calculator. These show you the calculation you have entered as well as the result.

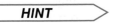 **HINT** offers some help for dealing with a problem. This might be a short cut, a way of remembering or a reminder that your VPAM can do the job.

Exam Questions means a question of the sort you may find in an examination. Try these! Practice makes perfect. We have given the numerical parts of the answers to these questions at the back of the book.

To know whether you need to learn all of the material in the book, you should consult the specification for the physics examinations you are taking. You will need most of the book's content, but some physics that is not in your specification may be used to illustrate the maths. You may want to leave out the mathematics of waves in Chapter 15.

About the calculator

We have assumed that you will use a VPAM calculator. All of the instructions are for the Casio *fx*-83WA which is representative of most. If you have a different calculator, find out from its manual how to do each sum.

Thanks

I wish to thank Beth Hutchins, John Hepburn and the other staff of Nelson Thornes for the guidance and help they have given to me during the preparation of this book.

John Rounce

Chapter 1

Multiplication and division

After completing this chapter you should:

- *be able to work out formulae that involve multiplications and divisions*
- *have used your calculator successfully for multiplications and divisions*
- *be aware of common mistakes made when numbers divide*
- *have learnt to make rough estimates.*

1.1 Multiplying

What is multiplication really?

We can use a little bit of physics to illustrate that multiplying or **multiplication** is an alternative to repeating a lot of adding.

A sodium chloride molecule is made up of one sodium atom and one chlorine atom. So there are two atoms in each molecule. The number of atoms in seven molecules is

$$2 + 2 + 2 + 2 + 2 + 2 + 2 = 14$$

The **equals sign** (=) means 'has the same value as' and seven twos give us 14. A simpler equation to write is $7 \times 2 = 14$ which reads as '7 times 2 equals 14' or '2 multiplied by 7 equals 14'.

Physics calculations for example would be ridiculously difficult without using multiplication.

The result of a multiplication is called its **product** and 14 is the product here. The term may also be used for the 7×2 that works out to the 14. The multiplying numbers, the 2 and the 7 are called **factors** of the product.

The number of atoms, 14, could be deduced in another way by saying that there are 7 sodium atoms and 7 chlorine atoms, i.e. $7 + 7$ atoms in all which is 2 times 7. So 2 times 7 equals 14 and 7 times 2 equals 14 or $7 \times 2 = 2 \times 7$. Adding is like this, $7 + 2 = 2 + 7$.

KEY FACT *The order of multiplication is not important.*

The importance of multiplication in physics calculations is illustrated by *Fig. 1* which shows just a few of the products we meet.

Now consider a common salt solution that contains 9 grams of salt in each litre. How many grams of salt will there be in 8 litres?
$9 \times 8 = 72$ gram
(The units for calculations are written in the singular (gram not grams) – see Chapter 5.)

The rule that the order of multiplication is not important can be used to deduce another important fact as follows:

$$30 \times 20 = 3 \times 10 \times 2 \times 10 = 3 \times 2 \times 10 \times 10 = 6 \times 100 = 600$$

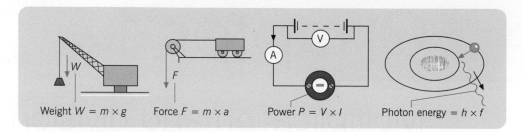

Fig. 1 *Some products met in physics.*

and shows how the zeros of the 30 and 20 can be neglected while the 3 and 2 of the 30 and 20 can be multiplied to give 6, the zeros being joined onto the 6 afterwards.

Similarly for 500×300 we have $5 \times 3 = 15$ and we join four zeros to this to get 150 000. In the same way $50 \times 3 = 150$.

KEY FACT   For 500×300 work out 5×3, join on 0000 to get 150 000.

Using an electronic calculator for multiplying

When numbers are to be multiplied that are not simple whole numbers you can use an electronic calculator. You may wish now to use the 9×8 example as a first exercise on the use of your calculator. More challenging examples will soon be suggested.

The *fx-83WA* calculator is switched on by pressing the key marked AC/ON and the OFF key is located top right on the key pad as shown in *Fig. 2*.

Fig. 2 *A suitable calculator for A level physics calculations.*

The keys to use for the multiplication are simply

$\boxed{9}\boxed{\times}\boxed{8}\boxed{=}$

You will see that the *fx-83WA* calculator shows the calculation it is doing and gives the answer on the right of the screen.

The answer you should expect is 72 but the calculator's screen may show it as 72. or 7.2^{01} or something like 72.0000. This is because the calculator has a number of different ways of working and displaying answers – it has a number of different **modes** of operation.

You can see the modes offered by your *fx-83WA* calculator by pressing the key marked MODE. Pressing it three times shows all the possibilities. *Scientific mode* is best for physics calculations and will be explained in Chapter 4 but until then COMP will be simpler to use. So press the mode key to get COMP, SD and REG displayed. You will need the mode for ordinary calculations (computations) rather than for statistics so you will select COMP by pressing 1 on the keypad.

Pressing the mode key again and a second time shows the choice between Deg, Rad or Gra. When you work with angles you will need the degree mode so select Deg by pressing 1.

Use the mode key again and press it until the choice is displayed of Fix, Sci or Norm. Press 3 to select the Norm mode. The choice then is between Norm1 and Norm 2. Choose Norm 2. This mode is suitable for all the exercises in Chapters 1 to 3. The answer given for our $9 \times 8 =$ calculation will, in this mode, be displayed as 72. The dot to the right of the 72 is a **decimal point** and will become important when decimal fractions are discussed in Chapter 2. Until then take no notice of it.

When your calculator is switched off it remembers the mode setting that you have chosen. When you switch the calculator on again it is a good idea to try a simple calculation like the $9 \times 8 =$ and see that you get the expected answer. You will in this way discover whether the calculator's batteries are working properly and confirm that you have the mode you want.

During any calculation if a wrong figure is entered the key marked DEL can be pressed to delete that last figure entered or the AC (AC/ON on the *fx-83WA*) can be used to clear whatever is on the screen.

Test Yourself   Exercise 1.1.1

1 Work out the following products to give a single number. Do this without your calculator and then check your result with your calculator.

(a) 5×3 (b) 323×2 (c) $4 \times 2 \times 7$ (d) $5 \times 3 \times 1$ (e) 4000×400

2 26 centimetres $\times 2 = ?$

Decimal numbers

In every-day life and in most of our physics work we write numbers in decimal form. So for example the number of days in a year is 365 and the 3, because of its position (third digit from the right), tells us how many hundreds of days, while the 6 means 6 tens and the 5 tells us the 5 ones (single days). So the positions (or *columns*) of this decimal number tell us the hundreds, the tens and the ones involved whereas the columns of a binary number tell us the number of ones, twos, fours, eights, etc. For example the binary number 1001 means an 8, no fours, no twos and a one. The decimal number is 9.

Moving the digits of a decimal number one place to the left and putting a zero in the right-hand position to show that this position is empty will change an 8 into 80 and we have 8 tens or 8×10 instead of 8.

KEY FACT *Multiplication of a whole number by 10 requires a zero to be fitted onto the number.*

Similarly $\times 100$ requires two extra zeros.

$9 \times 100 = 900$

For a more difficult multiplication consider 90×70.

Until electronic calculators became common, products were listed in *multiplication tables*. (See *Fig. 3*) which listed the products of numbers from '1 times 2' to '12 times 12'.

These tables had to be memorised. For 90×70, you would know that 9 times 7 is 63 and then join on two zeros to get 6300.

If you don't know that nine sevens is 63 and you don't have a calculator then you'll have to add nine sevens or say ten sevens is 70 and knock off one 7 to get 63. Calculators can now be used instead. If you do remember results from tables it will help with rough checks which are described below.

$1 \times 1 = 1$	$1 \times 2 = 2$	$1 \times 3 = 3$	$1 \times 4 = 4$
$2 \times 1 = 2$	$2 \times 2 = 4$	$2 \times 3 = 6$	$2 \times 4 = 8$
$3 \times 1 = 3$	$3 \times 2 = 6$	$3 \times 3 = 9$	$3 \times 4 = 12$
$4 \times 1 = 4$	$4 \times 2 = 8$	$4 \times 3 = 12$	$4 \times 4 = 16$
$5 \times 1 = 5$	$5 \times 2 = 10$	$5 \times 3 = 15$	5×4
$6 \times 1 = 6$	$6 \times 2 = 12$	$6 \times 3 = 18$	6
$7 \times 1 = 7$	$7 \times 2 = 14$	$7 \times 3 = 21$	
$8 \times 1 = 8$	$8 \times 2 = 16$	$8 \times 3 = 24$	
$9 \times 1 = 9$	$9 \times 2 = 18$	$9 \times 3 = 27$	
	$10 \times 2 = 20$		

Fig. 3 *Examples of multiplication tables.*

If you do memorise the products of numbers up to 12 for example, it will be easy for you to make rough checks of your calculations as explained below.

Test Yourself

Exercise 1.1.2

Write down the single number for each of the following products. (Don't use a calculator.)

1 (a) 7×10 (b) 8×100 (c) $5 \times 10 \times 10$ (d) 9000×100 (e) $100 \times 100 \times 100$

Rough checks

A calculator is essential for your physics work to deal with awkward numbers and so it may be thought that using it for simpler numbers as well will make tables completely unnecessary. This is not the case.

When using a calculator for any multiplication or other mathematical operation it is very easy to press the wrong key without noticing the error. It is also easy to accidentally repeat an entry. So you should always get a rough idea, an estimate, of the result that the calculator should give.

For example the product of 21 × 59 which is not very different from 20 × 60 is going to be close to 1200. In fact it equals 1239.

21 × 59, estimate 20 × 60

A moment of distraction might have caused the × 59 entry to be repeated so that 21 × 59 × 59 gives 73 101; but the rough estimate would detect the error.

Ideally a rough check is done without your calculator. This avoids your getting an estimate that is affected by the same mistake that it should be detecting. This is where remembering multiplication table results is useful. Multiplying for rough checks can also be quicker when you are using results you remember from tables rather than using your calculator.

In the above example the estimate was obtained using numbers, 20 and 60, that ended in zeros. This made the estimation very easy. As suggested already the product can be worked out by multiplying the 2 by 6 to get 12 and then two zeros are attached to give 1200.

Test Yourself

For each of the following multiplications estimate the product and select its exact value. (Don't use your calculator.)

1 21 × 49

| **A** 1029 | **B** 10 290 | **C** 103 | **D** 9021 |

2 35 × 2 × 41

| **A** 287 | **B** 2870 | **C** 28 700 | **D** 35 241 |

3 105 × 32

| **A** 10 532 | **B** 32 105 | **C** 33 600 | **D** 3360 |

A physics calculation with a rough check

The heat produced in an electric heater can be calculated from the formula

heat produced (in joules) = $V \times I \times t$

A formula tells you how to work out the answer you want. Its left-hand side states the quantity to be worked out, the right-hand side contains symbols for which values are put in. Here the formula requires three numbers to be multiplied. The V is the abbreviation for voltage or potential difference, I represents the value of the current flowing through the heater (measured in amperes) and t is the time in seconds for which the heater is used. We are to calculate the heat produced if $V = 6$ volts, I is 2 amperes and t is 330 seconds.

Note that it is usual to omit multiplying signs as long as the omissions cause no confusion. Consequently the formula is normally written as $V I t$. However the × signs have to be included when the symbols V, I and t are replaced by figures.

Heat produced = $V I t = 6 \times 2 \times 330 = 12 \times 330 = 3960$ joule
For a rough check 12 × 300 is 12 times 3 with 00 following which is 3600.

It will be explained later (Chapters 4 and 5) that this answer would be better written as 4×10^3 joule or 4×10^3 J.

Why not repeat this calculation with the figures entered into your calculator in a different order? This way you can see that the order of multiplying numbers is not important.

Test Yourself

Exercise 1.1.4

1 Calculate the weight W in newtons given by the formula $W = mg$ for $m = 30$ kilograms and $g = 10$ newtons per kilogram.

HINT $\quad W = m \times g$

2 The pressure experienced by a liquid at a depth h is greater than the pressure at the surface by an amount P given by P $= h\rho g$. The *Greek letter* ρ (pronounced *roe*) denotes the density of the liquid and g is the gravitational intensity or acceleration due to gravity. With h in metres, ρ in kilograms per cubic metre and g in newtons per kilogram (or metre per second per second) P is calculated in pascals (Pa). Determine P for $h = 2$, $\rho = 1000$ and $g = 10$.

1.2 Division

Division? What is it?

Division or dividing is often known as **sharing**.

Consider 12 litres of liquid contained in a very large bottle. It is poured out into four flasks with equal amounts in each. The liquid is *shared* (equally) by the four flasks or *divided* (equally) between them. The result of the division is 3 litres (in each flask) and we write

12 divided or shared by 4 = 3 or 12 ÷ 4 = 3 or 12/4 = 3 or $\frac{12}{4} = 3$

On a calculator

$\boxed{1}\,\boxed{2}\,\boxed{\div}\,\boxed{4}\,\boxed{=}$

are the keys to use. For 12/4 or $\frac{12}{4}$ we can say '12 over 4' instead of 12 divided by 4.

Dividing is the reverse of multiplying as seen by multiplying the 3 by 4 to get back to the original 12 in our example. When division occurs we should picture the sharing (into 4 portions in our example) and think of one portion being retained as our answer while the other shares can be forgotten. The 12 has been reduced to 3.

You are unlikely to hear the term 'shared' used in an A level physics or mathematics course, *dividing* is preferred.

The expression 'divided *into*' is deliberately avoided in this book. It is common to hear 'divided into two' meaning halved so that 6 divided into 2 might be interpreted as 3 but also one hears of '6 divided into 2' when the meaning is $\frac{2}{6}$ and this of course equals a third! In practice it is usually obvious which meaning is implied but take care.

The *Test Yourself* exercise below contains simple physics calculations involving dividing. You should try these.

1 The speed or velocity v of an object that travels steadily for a time t and covers a distance d is given by $v = \frac{d}{t}$. If t is measured in seconds, d in metres then v is in metres per second. Using these units calculate v for $d = 36$ and $t = 12$.

2 The formula $v = u + at$ relates the initial velocity u of an object, the final velocity v after a time t and a constant value of acceleration a. The equation can be rearranged (see Chapter 8) to give a formula for t which is $t = \frac{v - u}{a}$. t will be in seconds if u and v are measured in metres per second and a is in metres per second per second. Using these units calculate t for $u = 31$, $v = 76$ and $a = 5$.

Order of division

If 24 is halved (divided by 2) and then divided by 4 we get 12 then 3. If instead the 24 is divided by 4 then by 2 the result is also 3. This illustrates the rule that the order of division is not important. A similar statement has been made about multiplication.

$$24 \div 2 \div 4 = 3 \qquad 24 \div 4 \div 2 = 6 \div 2 = 3$$

For division a more important fact is that two dividing numbers, the 2 and the 4 here, can be multiplied together and their product, namely 8, can be used to divide into the 24.

$$\frac{24}{2} \div 4 = \frac{24}{2 \times 4} = \frac{24}{8} = 3$$

Note too that $24 \div 6 \div 2$ means first divide the 24 by 6 and *then* divide by 2. The answer is 2. If the 6 were divided by 2 to give 3 and then the 24 were divided by this an incorrect answer of 8 would result. It is better to write $\dfrac{24}{6 \times 2}$ rather than $24 \div 6 \div 2$ to avoid this.

Another possible source of confusion is an expression written as $24 / 12 / 2$. This could be interpreted as $24 / 12$ (which is 2) then divided by 2 to give 1 *or* the denominator of $12 / 2$ could be worked out as 6 so that 24 divided by 6 gives 4. The answer to this problem is to avoid having more than one dividing line. (See page 22.)

Dividing by 10, 100, etc

Consider a number ending in one or more zeros, for instance 300. If this is divided by 10 a zero is removed from the right-hand end of the 300 to leave 30.

$$\frac{300}{10} = 30$$

Dividing the 300 by 100 requires two zeros to be removed. These facts can be very useful for rough estimates.

$$300 \div 100 = 3 \qquad \text{(Two zeros have been removed from the 300.)}$$

What about dividing by 1?

Bearing in mind that sharing is the same as dividing we can say that sharing a number (of items) between *not* 2 or more, but just one person would mean that the one person gets all the items. So the result of 7 divided by 1 is 7 for example.

If the letter x is used to mean 'the number in which we are interested' then

$$\frac{x}{1} = x$$

Write down the result of each of the following divisions:

1 (a) 40 / 10   (b) 7000 / 100   (c) $\dfrac{400}{2 \times 100}$ (d) 400 / 200   (e) $\dfrac{8 \times 300}{30 \times 2}$ (f) $\dfrac{8 \times 500}{40 \times 20}$

(g) $\dfrac{24 + 6}{11 + 19}$ (h) $\dfrac{20 + 6}{4 \times 9}$

Using your calculator for division

As with multiplying it is worth testing your calculator with two simple numbers, for example dividing 6 by 3. You know the answer to be 2 and using the calculator keys the sequence

$$\boxed{\text{AC}}\ \boxed{6}\ \boxed{\div}\ \boxed{3}\ \boxed{=}$$

will give the correct answer.

Next try dividing bigger numbers, $\frac{147}{21}$ perhaps, and for these figures the answer is 7. As suggested earlier a rough estimate of the answer is a useful check on the calculator's answer. Well $\frac{140}{20}$ equals 7.

Work out each of the following divisions to get a single number answer, use your calculator:

1 221/17   **2** $\frac{301}{43}$ **3** 135 kilometres divided by 27 hours (Answer in km per hour.)

Numbers with decimal points

Dividing numbers using a calculator often produces an answer that contains a decimal point. Dividing 11 by 5 gives 2.2 for example and is read as two point two. It will not be necessary for you to understand such numbers until they are discussed in Chapter 2.

Mistakes to avoid when dividing

Any combination of numbers and signs can be called an **expression**.

A mistake can easily be made when working out an expression like $\dfrac{32 + 180}{4}$. The 4 is dividing into the whole of the numerator. Add the 32 to the 180 to get 212 and then divide this by the 4 to get the correct answer of 53. Dividing the 32 by 4 to get 8 and then adding this to 180 would give the wrong answer. This would however be the correct procedure if the expression had been $\frac{32}{4} + 180$ where the 4 is only under the 32.

Another kind of mistake can be made because of the way calculators work

We want to calculate $\dfrac{390 \times 470}{390 + 470}$. A rough estimate for this is $\dfrac{400 \times 500}{900}$ or $\dfrac{200\,000}{1000}$.

So the answer is about 200.

Using the calculator it is very easy to get a wrong answer of 940 by using the key sequence

$$\boxed{3}\,\boxed{9}\,\boxed{0}\,\boxed{\times}\,\boxed{4}\,\boxed{7}\,\boxed{0}\,\boxed{\div}\,\boxed{3}\,\boxed{9}\,\boxed{0}\,\boxed{+}\,\boxed{4}\,\boxed{7}\,\boxed{0}\,\boxed{=}$$

In this case the calculator has worked out $\dfrac{390 \times 470}{390} + 470$.

A simple way of avoiding this error is to first add the 390 and 470 to get 860 and write this figure down somewhere. Then after clearing the calculator use it to evaluate $390 \times 470 \div 860$ and get the correct answer of 213 ohms.

KEY FACT *Do the adding for the denominator first.*

An expression like the one worked out here can be met in a calculation of the resistance of two electrical conductors connected in parallel as in *Fig. 4a*. The opposition to the electric current, i.e. the resistance due to R_1 and R_2 in parallel is the same as a single resistance of size R, as in *Fig. 4b*. The formula for R is $\dfrac{R_1 \times R_2}{R_1 + R_2}$ and if for example R_1 is 390 ohms and R_2 is 470 ohms then $R = \dfrac{390 \times 470}{390 + 470}$.

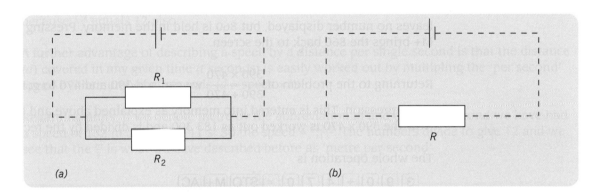

Fig. 4 *(a) Resistances in parallel. (b) Equivalent resistance.*

A neater way of dealing with this last problem is to make use of the calculator's memory. This technique is discussed below. Making use of a calculator's reciprocal function is discussed in Chapter 2 and the use of brackets will be met in Chapter 3. These procedures provide alternative answers to the problem.

Here is a similar difficulty that can arise when a calculator is used for dividing. Consider the expression $\dfrac{9603}{11 \times 9}$. It must approximately equal $\dfrac{10\,000}{10 \times 10}$ or 100. When using a calculator it is essential to realise that the 9603 is divided by 11 and then *divided* by 9 or alternatively is divided by 99 to give an answer of 97.

An expression like $\dfrac{97 + 23}{40}$ is easily handled if the 97 and 23 are added and then the equals key is pressed to give 120 which is then divided by 40. This procedure avoids getting $97 + \frac{23}{40}$.

Example

How many oxygen molecules are there in 160 grams if 32 grams contains 600 000 million million million? (Written more easily as 6×10^{23}, see Chapter 4.)

Answer

How many lots of 32 are there in 160?

The number of 32s in 160 is 160 divided by 32, which equals 5.

So the number of oxygen molecules in the 160 grams is 5 times the 6×10^{23} and is 3 000 000 million million million or 3 million million million million.

(How would you use a calculator for such a multiplication? All is explained later.)

Test Yourself

Exercise 1.2.7

1 How many twelves are there in 72?

2 How many 3 metre lengths of wire can be cut from a 36 metre length?

3 How many thirteens are there in 741?

> **HINT** *Use a calculator.*

Exam Questions

Exam type questions to test understanding of Chapter 1

Exercise 1.2.8

1 A load of 4 N hangs at the lower end of a vertical, spiral spring. The spring constant is 8 N per m. The elastic energy, in J, in the system is
   A 32
   B 1
   C 64
   D 128

> **HINT** *The 4 N is the force F pulling on the spring and producing an extension e so that the work done or energy stored is $E = \frac{1}{2}F \times e$. The spring constant k equals $\frac{F}{e}$ so $E = \frac{1}{2}F \times \frac{F}{k}$. Force means push or pull.*

2

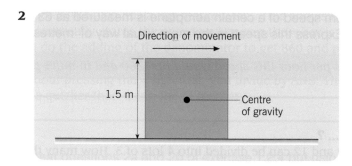

Fig. 5

A uniform square block is sliding with uniform speed along a rough surface as shown in *Fig. 5*. The force used to move the block is 200 N. The moment of the frictional force acting on the block about the centre of gravity of the block is

A 150 N m, clockwise   **B** 150 N m, anticlockwise

C 300 N m, clockwise   **D** 300 N m, anticlockwise

(AQA 2000)

HINT *The frictional force F, is equal to the applied force and is oppositely directed because the speed is constant. The moment or turning effect of F about the centre = F × 1.5/2.*

3 (a) State what is meant by an electric current.

   (b) A car battery has a capacity of 40 ampere-hours. That is, it can deliver a current of 1.0 A for 40 hours. When delivering the current, the potential difference across the terminals of the battery is 12 V. Calculate the electrical energy available from the battery.

(OCR 1999, part question)

HINT *Potential difference × current × time in seconds gives energy in joules. 1 hour = 3600 seconds.*

4 The capacitive reactance X_c of a capacitor is given by $1 / 2\pi f C$. Calculate an approximate value for X_c given that π is approximately 3, $f = 100$ and $C = 2/10\,000$. The units are hertz for f, farad for C and ohm for X_c.

HINT *The $1 / 2\pi f C$ means $\dfrac{1}{2 \times \pi \times f \times C}$.*

5 A certain light bulb is used with a 12 volt supply and the resulting current is 2 amperes. Calculate: (a) its power and   (b) its resistance.

HINT *Power is the energy per second and equals voltage × current. Resistance is voltage ÷ current. Working in volts and amperes, power will be in watts and resistance in ohms.*

Answers to Test Yourself Questions

Exercise 1.1.1
1 (a) 15   (b) 646   (c) 56   (d) 15   (e) 1 600 000
2 52 centimetre

Exercises 1.1.2
1 (a) 70   (b) 800   (c) 500   (d) 900 000
   (e) 1 000 000

Exercise 1.1.3
1 A
2 B
3 D

Exercise 1.1.4
1 300 N
2 20 000 Pa

Exercise 1.2.1
1 3 metre per second
2 9 second

Exercise 1.2.2
1 (a) 4   (b) 70   (c) 2   (d) 2   (e) 40   (f) 5   (g) 1
   (h) 1

Exercise 1.2.3
1 13
2 7
3 5 kilometre per hour

Exercise 1.2.4
1 (a) 20   (b) 34   (c) 4
2 (a) 3   (b) 5   (c) 4   (d) 2
3 95 watt

Exercise 1.2.5
See answers to exercise 1.2.4.

Exercise 1.2.6
1 20 pence per transistor or £0.20 per transistor
2 21 000 metre per minute

Exercise 1.2.7
1 6
2 12
3 57

Chapter 2

Fractions

After completing this chapter you should:

- *have checked your knowledge of fractions*
- *learnt to simplify expressions, e.g. by cancelling*
- *be able to work out combinations of fractions, reciprocals and other numbers*
- *be familiar with decimal fractions*
- *be able to work out percentages.*

2.1 Fractions and their names

What is a fraction?

The resistance (R) of two resistors (R_1 and R_2) connected in parallel can be calculated from the equation $\dfrac{1}{R} = \dfrac{1}{R_1} + \dfrac{1}{R_2}$ and all three terms involved are fractions. How do you work out R when values are given for the other two resistances? Let's look at fractions.

In everyday life a fraction is part of a whole thing. A half is the best example. The whole thing could be a whole litre of liquid and a half of it is obtained by separating or **dividing** the one litre to get two equal parts and retaining one of them. Because two portions are created to get a half we say that we have 'divided by 2'. Sometimes the term *shared by* is used instead of *divided by* and this is appropriate if we imagine the litre of liquid shared (equally) by two beakers.

> **KEY FACT** A half may be written as 1/2 or $\frac{1}{2}$ which means 1 divided by 2.

The top number of a fraction is called the **numerator** and the number dividing it is the **denominator**. In mathematics, as opposed to everyday life, the term **fraction** is in fact used for *any* number divided by another so $\frac{1}{2}$ is a fraction but so is $\frac{7}{2}$. When a fraction is greater than one (its numerator greater than the denominator) it is called an **improper fraction**. Otherwise it is a **proper fraction**.

The example used above started with a litre of liquid. If instead a pint were used the half would of course be a smaller quantity of liquid. So we use the expressions 'half *of* a litre' or 'half *of* a pint', half litre or $\frac{1}{2}$ pint.

In everyday life we may refer to a 'two litre bottle' rather than a 'two litres bottle'. When Chapter 5 of this book is reached we will prefer '2 litre' rather than the plural 'litres'. So for example $\frac{1}{2}$ litre × 12 = 6 litre.

The fraction $\frac{1}{6}$ is described as a *sixth*, $\frac{1}{8}$ as an *eighth* and so on but $\frac{1}{2}$ and $\frac{1}{4}$ alone have special names, a *half* and a *quarter* of course. A fraction can be quoted as one number *over* another, e.g. $\frac{5}{8}$ may be read as '5 over 8'.

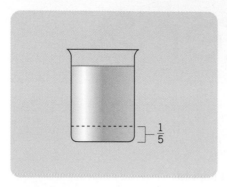

Fig. 1 *A fraction*.

What is meant by $^3/_4$? Consider three litres of liquid occupying and filling three 1 litre beakers. Since dividing by 4 means sharing between 4 so we could divide the liquid equally between 4 new beakers and each then has $\frac{3}{4}$ of the liquid. Another way of getting the same result is illustrated in *Fig. 2*.

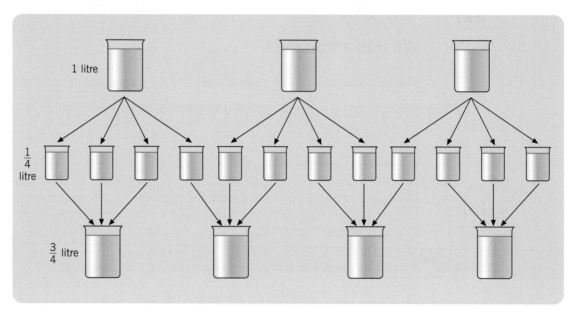

Fig. 2 *Three-quarters*.

We could share each litre equally between four small beakers so that each contains a quarter of a litre. Finally three of these can be put together in each new beaker. The liquid in each new beaker is three times a quarter and is the same as three litres divided between four.

$\frac{3 \; litres}{4}$ or $\frac{3}{4}$ litres = $\frac{1}{4}$ of a litre $\times 3$ or 'three quarters'

KEY FACT $\frac{3}{4} = \frac{1}{4} \times 3$.

$\frac{3}{4}$ is seen more often than $\frac{2}{4}$ because two quarters make a half and therefore $\frac{2}{4}$ is usually replaced by $\frac{1}{2}$. This is simpler and is preferred because it employs smaller numbers.

Test Yourself

1 (a) How much do you get in each container if:
    (i) 3 kilograms is shared between 4 containers
    (ii) 3 litres is shared by 8
    (iii) 4 litres is shared by 12?
  (b) What fraction results if 4 is divided by 12

> **HINT** *Shared between* or *shared by* means *divided by.*

2 Express each of the following fractions in words:

  (a) 1/2   (b) 1/6   (c) 1/4 litre   (d) 1/100

3 Write as a fraction:

  (a) a fiftieth   (b) seven over 3   (c) a twelfth   (d) an eleventh

4 There are two fours in 9. How many fours are there in:

  (a) 22   (b) 43   (c) 153

> **HINT** *Divide number by 4.*

2.2 Simplifying expressions

Any combination of numbers can be called an **expression**, for example. $\frac{9+11}{3+2}$ or *nAve* which features in the formula *I = nAve*.

> **KEY FACT** *Simplifying an expression means rewriting it with smaller numbers or with fewer numbers and symbols.*

It is simplifying with equations containing symbols that is most important but the $\frac{9+11}{3+2}$ example will be helpful and in this chapter we are interested in fractions. The fraction simplifies to $\frac{20}{5}$ when the addings are done and the $\frac{20}{5}$ (which for the moment we will say is 20 shared by 5) simplifies to give the value of the expression as 4.

Simplifications like this where only simple numbers are involved can be quicker and safer than using a calculator.

You can often simplify a fraction by dividing top and bottom of the fraction by a suitably chosen number. The $\frac{20}{5}$ above became 4 by dividing top and bottom by 5 to give $\frac{4}{1}$ which is simply 4. Multiplying top and bottom of $\frac{3/5}{4/5}$ by 5 simplifies it to $\frac{3}{4}$. A better example would involve decimal fractions which are discussed later.

The rule that is being used is

> **RULE** *The value of a fraction is not changed if its numerator and denominator are multiplied by the same number or divided by the same number.*

The rule is illustrated well by the fact that two quarters is a half

$$2 \times \tfrac{1}{4} = \tfrac{2}{4} = \tfrac{1}{2}$$

so that $\tfrac{2}{4} = \tfrac{1}{2}$ and we can say that top and bottom of the $\tfrac{1}{2}$ have been multiplied by 2. Changing $\tfrac{2}{4}$ to $\tfrac{1}{2}$ involves dividing top and bottom by 2. Similarly a half is the same as three sixths and $\tfrac{1}{2}$ becomes $\tfrac{3}{6}$ by multiplying top and bottom by 3.

A fraction will, whenever possible be written using the smallest figures. So for example $\tfrac{8}{16}$ is preferably stated as $\tfrac{1}{2}$.

As regards simplifying a formula suppose we consider $I = nAve$ where I (perhaps surprisingly) represents current. Suppose that you want a formula not for current but for 'current per unit area' which is $\tfrac{I}{A}$. The formula for this is $\tfrac{nAve}{A}$. Now we divide top and bottom by A and get nve.

As a further example of simplifying look at $\tfrac{42}{18}$. Seeing that 2 divides into both 42 and 18, we get $\tfrac{21}{9}$ and dividing top and bottom by 3 gives $\tfrac{7}{3}$.

The fraction $\tfrac{42}{18}$ has been simplified with the 42 and 18 being replaced by smaller numbers.

The simplifying just described can be called **cancelling** because

KEY FACT   *Cancelling means removing a number or replacing it by a smaller number.*

KEY FACT   *Cancelling in a fraction is often achieved by dividing numerator and denominator by a suitably chosen number.*

Simplifying a fraction by adding or subtracting numbers like changing $\tfrac{3+2}{7}$ to $\tfrac{5}{7}$ is however not regarded as cancelling.

A common way of displaying the cancelling of the $\tfrac{42}{18}$ is crossing out the cancelled figures and marking in the figures replacing them as shown in *Fig. 3*.

There is however the risk when cancelling of creating a confusion of figures so that the simplification cannot be understood, as in *Fig. 4*.

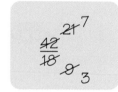

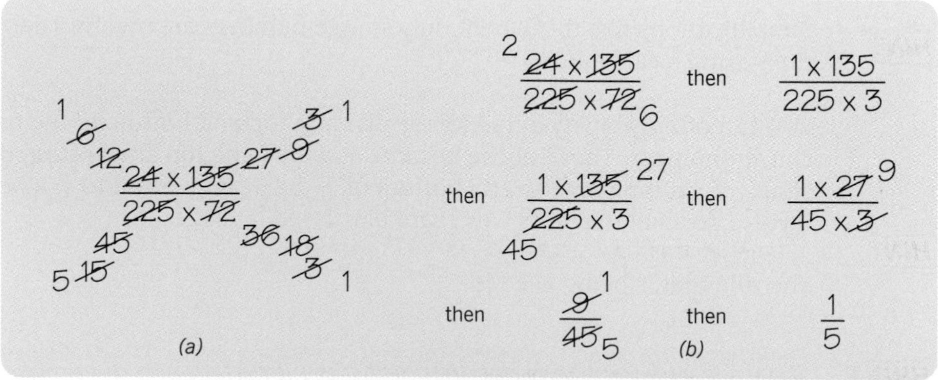

Fig. 3 *Cancelling.*

Fig. 4 *Cancelling to confusion. (a) Avoid this. (b) Better.*

As seen in *Fig. 4* it would be quicker to use a calculator in this case (and as a rough check

$$\frac{20 \times 140}{200 \times 70} = \frac{1 \times 2}{10 \times 1} = \frac{2}{10} = \frac{1}{5}.)$$

In this last equation where the ones cancelled the $\frac{1 \times 2}{10 \times 1}$ could be written as $\frac{2}{10}$ by realising that $1 \times 2 = 2$ and $10 \times 1 = 1$.

KEY FACT   *A 1 multiplying a number can be left out.*

It was also pointed out in Chapter 1 that a one dividing a number can be omitted.

KEY FACT   *A 1 dividing a number can be omitted.*

Note however that the 1 obtained when $\frac{2}{10}$ was cancelled to give to give $\frac{1}{5}$ could not be left out. It was not multiplying or dividing a number. We will meet cancelling again in Chapter 8.

Now consider multiplying a number, say 5, by 3 and then dividing by 3.

$5 \times 3 = 15$ and dividing the 15 by 3 gives 5 which is the same as we began with.

KEY FACT   *Multiplying and dividing a fraction by the same number have opposite effects. The two processes cancel.*

This effect can be seen in the expression $5 \times \frac{3}{3}$ where cancelling in the fraction gives $5 \times \frac{1}{1}$ or 5×1 or just 5 and agrees with multiplying the 5 by 3 having the opposite effect to dividing it by 3.

Test Yourself	Exercise 2.2.1

1 Simplify each of the following expressions:

(a) $\dfrac{1/2}{1/4}$ (b) $\dfrac{3/8}{4/8}$ (c) $\dfrac{23/3}{7/6}$ (d) $\dfrac{43/100}{6/10}$

2 Simplify each of the following fractions:

(a) $\frac{4}{8}$

> **HINT**   *Divide top and bottom by 2.*

(b) $\frac{12}{3}$ (c) $\frac{100}{10}$ (d) $\frac{200}{40}$ (e) $\frac{40}{200}$ (f) $\frac{240 \text{ volts}}{20}$

(g) $\dfrac{2 \times 7}{7 \times 3}$

> **HINT**   *The sevens cancel, i.e. divide top and bottom by 7.*

(h) $\frac{300\,000}{20\,000}$

> **HINT**   *Divide top and bottom by 100 and repeat this.*

A common mistake when cancelling

If you meet a fraction such as $\frac{24+23}{4}$ and realise that 4 divides into 24 you might make the mistake of cancelling the 4 with the 24 to give 6 and then an incorrect 6 + 23 results. Here it is important to remember that any cancelling we are doing is really the process of dividing the fraction's top (the *whole* top) and its bottom by the same chosen number. Since the 23 will not divide nicely by 4 you can't cancel. Some simplification is obtained when the 23 is added to the 24 to give 47 and the simpler fraction $\frac{47}{4}$ is left. This too cannot be cancelled.

The fraction $\frac{24+32}{4}$ could be simplified by cancelling the 4 with both the 24 and the 32 to give 6 + 8 and the correct answer of 14. (Of course the alternative here is to add the 24 and 32, get 56 and cancel this with the 4 to get $\frac{28}{2}$ and then $\frac{14}{1}$ or 14 as expected.)

Sometimes a similar situation arises but involves an algebraic expression and does not allow the adding of the parts of the numerator. $\frac{6x+8}{2}$ is an example. (You should regard the x as an abbreviation for 'number not yet known' or an 'unknown quantity' and remember that $6x$ means 6 times the x number.)

The addition cannot be done until the number x is known but some cancelling is possible before then by dividing top and bottom by 2 to give $3x + 4$.

Example
As shown in *Fig. 5*, a 120 Ω resistor and a 30 Ω resistor are in series with a 12 V supply, the internal resistance of which can be neglected. Calculate the current that flows.

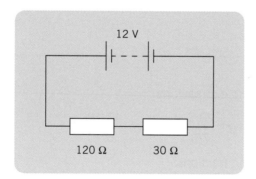

Fig. 5

Answer

The current measured in amperes and denoted by I is given by $I = \dfrac{12}{120+30}$.

So $I = \frac{12}{150} = \frac{4}{50} = \frac{8}{100} = 0.08$ ampere.

Alternatively we could write

$$I = \frac{12}{120+30} = \frac{4}{40+10} = \frac{4}{50} = \frac{8}{100} = 0.08 \text{ ampere,}$$

but $I = \dfrac{12}{120+30} = \dfrac{1}{10+30}$ is *wrong* because the cancelling by 12 is incorrect.

Test Yourself

Simplify each of the following expressions as far as possible but, just for this exercise, without adding or subtracting any numbers.

(a) $\dfrac{4+8}{2}$

HINT *Divide top and bottom by 2.*

(b) $\dfrac{3+5}{4}$ (c) $\dfrac{8+3}{2}$ (d) $\dfrac{16+6}{4}$ (e) $\dfrac{2}{4+4}$ (f) $\dfrac{3}{12+2}$

(g) $\dfrac{15}{10+5}$ (h) $\dfrac{12}{8+6}$ (i) $\dfrac{100}{10+10}$ (j) $\dfrac{120+30}{20+12}$

HINT *Divide top and bottom by 2.*

(k) $\dfrac{8-1}{4}$ (l) $\dfrac{5}{25-8}$ (m) $\dfrac{12x+21}{3}$

HINT *Divide top and bottom by 3.*

(n) $\dfrac{2x-6}{6}$ (o) $\dfrac{5x+3x}{4x}$

HINT *Divide top and bottom by x.*

2.3 Combining fractions

Multiplying and dividing fractions

Fig. 2 showed that $3 \times \frac{1}{4} = \frac{3}{4}$. Changing the order of multiplication gives $\frac{1}{4} \times 3 = \frac{3}{4}$. The effect of multiplying the fraction by 3 is to multiply just the numerator.

$$\frac{1}{4} \times 3 = \frac{1 \times 3}{4} = \frac{3}{4}$$

KEY FACT *A number multiplying a fraction multiplies just the numerator.*

If we write the $\frac{1}{4} \times 3$ as $\frac{2}{8} \times 3$ then our rule correctly gives $\dfrac{2 \times 3}{8}$ which is $\frac{6}{8}$ or the expected $\frac{3}{4}$.

Similarly

$3 \times \frac{3}{16} = \frac{3 \times 3}{16} = \frac{9}{16}$

(It is preferable to write fractions with a horizontal dividing line because a product like $\frac{2}{9} \times 4$ (which equals $\frac{8}{9}$) if written as 2/9 × 4 could be interpreted as 2 divided by 36 (which equals $\frac{1}{18}$). Beware of this confusion.)

We can also meet a fraction multiplied by another fraction.

An example here is $\frac{1}{2} \times \frac{2}{3}$ (half × two thirds) which clearly equals $\frac{1}{3}$ (one third) and agrees with the $\frac{1 \times 2}{2 \times 3}$ given by the rule below because this equals $\frac{2}{6}$ which is $\frac{1}{3}$.

| **RULE** | *To multiply two fractions multiply their numerators and multiply their denominators.* |

What is the effect of dividing a fraction by a number?

We know that 1 divided by 2 equals $\frac{1}{2}$ and dividing this by 2 gives a quarter, $\frac{1}{4}$. So $\frac{1}{2}$ divided by 2 equals $\frac{1}{2 \times 2}$ and the division has required the dividing number to multiply the denominator of the fraction. Similarly $\frac{2}{3} \div 5 = \frac{2}{3 \times 5}$.

| **KEY FACT** | *To divide a fraction, multiply the denominator.* |

A quarter divided by 2 can be written as $\frac{1}{4 \times 2}$ (and equals $\frac{1}{8}$). No immediate problem arises with $\frac{1/4}{2}$. However, 1/4/2 could be mistaken for 1 divided by 4/2 which means 1 divided by 2 and a half would be obtained instead of an eighth. This problem was mentioned in Chapter 1. Brackets are explained in Chapter 3 and we could write (1/4)/2 or $(\frac{1}{4})/2$ but it is best to avoid writing a fraction with more than one dividing line. So $\frac{1}{4 \times 2}$ is preferred to 1/4/2.

Often it will be necessary to simplify a fraction that is divided by a fraction. Consider the fraction $\frac{3/4}{7/8}$. The denominator ($\frac{7}{8}$) can be changed so that it is no longer a fraction if it is multiplied by 8. If this is done then the numerator (the $\frac{3}{4}$) must also be multiplied by 8 in order that the value of the fraction is not changed. The result is

$$\frac{8 \times 3/4}{8 \times 7/8} \text{ or } \frac{24/4}{7} \text{ or } \frac{6}{7}$$

There is however another way to get this result. The bottom fraction is removed, turned upside down or **inverted** and then is used to multiply the top fraction.

$$\frac{3/4}{7/8} = \frac{3}{4} \times \frac{8}{7} = \frac{3 \times 8}{4 \times 7} = \frac{24}{28} = \frac{6}{7}$$

| **RULE** | *So the rule is invert the denominator and multiply.* |

In the case of the $\frac{3/4}{7/8}$ you may say that simplification is unnecessary because, with the aid of a calculator you could work out that the fraction equals 0.857 (a decimal answer as explained later in this chapter) but a fraction involving algebra such as $\frac{a/b}{c/d}$ will be simplified to $\frac{ad}{bc}$.

Example
A vehicle moves very slowly but steadily and covers 11 metres in a time of 50 seconds.
Calculate the distance it would travel in 5 minutes (300 seconds) if it maintains the same speed.

Answer

Speed = distance divided by time taken = $\frac{11}{50}$ metre per second.

Distance covered (in 300 s) = speed × time = $\frac{11}{50}$ × 300 and multiplying the numerator by the 300 gives the distance as $\frac{3300}{50}$.

If you use a calculator now to divide 3300 by 50 the answer is 66 metres.

Example

Express a time of 7200 seconds in hours.

Answer

7200 s is changed to minutes by dividing the 7200 by 60 to give $\frac{7200}{60}$ to express this time in hours requires again dividing by 60 so the number of hours is $\frac{7200}{60}$ divided by 60.

To divide the fraction by 60 the dividing 60 is made to multiply the denominator and so the time is $\dfrac{7200}{60 \times 60} = \dfrac{7200}{3600}$ and, if you now use a calculator to divide 7200 by 3600 or simplify the fraction as explained below, the answer becomes 2 hours.

Example

The activity of a certain radioactive substance becomes halved after a time of 1 hour. To what fraction is it reduced after: (a) 2 hours   (b) 4 hours?

Answer

(a)   After 2 hours it will be halved twice. It will equal $\dfrac{original\ activity}{2} \div 2$ or $\dfrac{original\ activity}{2 \times 2}$.

So the answer is a quarter.

(b)   After 4 hours the fraction will be $\dfrac{original\ activity}{2 \times 2 \times 2 \times 2}$. The answer is a sixteenth.

A fraction *of*

Consider the meaning of '$\frac{1}{2}$ of 6 litres'. You will, from your everyday experience say correctly that this equals 3 litres.

KEY FACT   *A **fraction of** means a **fraction times** because **of** means **times**.*

We can regard **of** as meaning **times** so that '$^1/_2$ of 6 litres' becomes $^1/_2$ times 6 and equals 3. This rule is particularly useful for a fraction like $\frac{3}{4}$ which does not have 1 as its numerator because $\frac{3}{4}$ of say 12 is $\frac{3}{4} \times 12$ so that

$$\tfrac{3}{4} \text{ of } 12 = \tfrac{3}{4} \times 12 = \tfrac{36}{4} = 9$$

Example

Fig. 6 shows a potential divider circuit which provides a variable potential difference between the terminals P and Q.

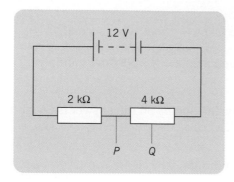

Fig. 6

The limits of the potential difference, in volts, obtainable between P and Q are:

A 0 to 2   **B** 0 to 4   **C** 0 to 8   **D** 2 to 12

Answer

(You may not understand this question but the answer shows an important use of fractions. The potential at a place and the potential difference between two places is measured in volts. The term *voltage* is often used for potential difference.)

Q can be moved to the P end of the 5 kΩ resistor to give zero volts between P and Q. The maximum potential difference is obtained with the whole of this resistor between P and Q. The fraction of the 12 V then obtained is equal to the fraction that 4 kΩ is out of the total of 4 kΩ + 2 kΩ or $\frac{4}{6}$. So the potential difference is $\frac{4}{6}$ of 12 or $\frac{48}{6}$. So this potential difference is 8 V and the answer to choose is C.

| **Test Yourself** | Exercise 2.3.1 |

1 Write each of the following as a single fraction:

   (a) $\frac{1}{3} \times 5$ (b) $\frac{1}{7} \times 10$ (c) $\frac{1}{9} \times 11$ (d) $\frac{2}{5} \times 7$

HINT ▷ *Multiply numerator only.*

2 Write each of the following as a single fraction:

   (a) $\frac{1}{4}$ of a volt ÷ 3   (b) $\frac{2}{5} \div 5$ (c) $\frac{23}{50} \div 2$

HINT ▷ *Multiply denominator by the dividing number.*

3 Write each of these products as a single fraction:

   (a) $\frac{1}{2} \times \frac{1}{4}$ (b) $\frac{1}{3} \times \frac{1}{40}$ (c) $\frac{2}{3} \times \frac{1}{5}$ (d) $\frac{3}{7} \times \frac{4}{5}$

HINT ▷ *Multiply numerators and multiply denominators.*

Reciprocals

If a fraction is turned upside down or inverted, the result is called the **reciprocal** of the original fraction. So the reciprocal of $\frac{2}{3}$ is $\frac{3}{2}$. We can also have the reciprocal of a number that is not a fraction. Consider the number 4 (which can be written as $\frac{4}{1}$). Its reciprocal is $\frac{1}{4}$.

KEY FACT *Inverting a fraction creates its reciprocal. The reciprocal of $\frac{2}{3}$ is $\frac{3}{2}$ and the reciprocal of x is $\frac{1}{x}$.*

On your calculator there is a key marked as x^{-1} (See *Fig 2* in Chapter 1). This key converts the number on the screen (x) into its reciprocal which is 1 divided by x or $1/x$. The reason why the x^{-1} symbol is often used instead of $1/x$ will be seen in Chapter 4.

So you could enter 1 divided by 4 ($=\frac{1}{4}$) into your calculator and expect its reciprocal to be 4. If you know that $\frac{1}{4}$ is the same as 0.25 you can enter this instead. The key sequence is then

$$\boxed{1}\ \boxed{\div}\ \boxed{4}\ \boxed{=}\ \boxed{x^{-1}}\ \boxed{=}$$

and you first get 0.25 and then 4.

You could repeat this last experiment (which gave the reciprocal of $\frac{1}{4}$) using other numbers in place of the 4.

The reciprocal key is useful for a parallel resistances calculations

using the equation $\dfrac{1}{R}=\dfrac{1}{R_1}+\dfrac{1}{R_2}$ where we might have $R_1 = 20$ and $R_2 = 5$ and want to determine R.

Then $\dfrac{1}{R}=\dfrac{1}{20}+\dfrac{1}{5}$. The x^{-1} key can be used to get $\frac{1}{20}$ equal to 0.05 and $\frac{1}{5}$ as 0.2.

So $\frac{1}{R} = 0.05 + 0.2 = 0.25$ and from your calculator R (= the reciprocal of 0.25) = 4.

The same procedure might be used for calculation of the focal length f, of a lens using the equation $\dfrac{1}{f}=\dfrac{1}{u}+\dfrac{1}{v}$ where u could be 5, v could be 20 and f works out to be 4.

You may recall from Chapter 1 that special care was needed when using a calculator to work out $\dfrac{420 \times 105}{420 + 105}$. Now we can consider another way of dealing with this calculation. We work out the inverse of the expression (i.e. we turn it upside down first) and then use keys

$$\boxed{4}\boxed{2}\boxed{0}\ \boxed{+}\ \boxed{1}\boxed{0}\boxed{5}\ \boxed{=}\ \boxed{\div}\ \boxed{4}\boxed{2}\boxed{0}\ \boxed{\div}\ \boxed{1}\boxed{0}\boxed{5}\ \boxed{=}$$

and get first 525 for the adding then, for our inverted expression, an answer of 0.01190 and when the reciprocal of this is obtained we get 84.

Adding and subtracting fractions

The above example involved the addition of the fractions $\frac{1}{20}$ and $\frac{1}{5}$ in the equation $\dfrac{1}{R}=\dfrac{1}{R_1}+\dfrac{1}{R_2}$.

We can combine the fractions in this equation before any values are given for the resistances. This will be shown in the example below but it will be worth looking first at an example with numbers.

$$\dfrac{1}{R}=\dfrac{1}{20}+\dfrac{1}{5}.$$

We won't work out the reciprocals of 20 and 5. Let's rewrite the two fractions so that they have equal denominators. Choose 20 for the new denominators. We get

$$\frac{1}{R} = \frac{1}{20} + \frac{4}{20}$$

which is one twentieth plus four twentieths and of course amounts to five twentieths. So $\frac{1}{R} = \frac{5}{20} = \frac{1}{4}$ and clearly $R = 4$.

$$\frac{1}{20} + \frac{1}{5} = \frac{1}{20} + \frac{4}{20} = \frac{5}{20} = \frac{1}{4}$$

Let's take as another example the expression $\frac{1}{3} + \frac{4}{5}$. For the new denominator choose 15 so that the expression is rewritten as $\frac{5}{15} + \frac{12}{15}$ and totalling the fifteenths gives 17 and the expression becomes $\frac{17}{15}$ for the answer.

But how do we choose a suitable number for the new denominator? A number is needed that is divisible by both of the original denominators. The 15 that was used when divided by 3 gives 5. The 3 divides into it exactly a whole number of times. The 15 also when divided by 5 gives a whole number, namely 3.

When 20 was used as a new denominator it was because both 20 and 5 divided into it. Alternatively 40 or 60 and many other numbers would have suited but why not use the simplest?

But suppose I don't easily think of a convenient number. The answer here is that the product of the original denominators will always be suitable even though it may not be the simplest, i.e. the smallest number that could be used. So for the $\frac{1}{20} + \frac{1}{5}$ you could multiply 20 by 5 to get 100 and use this as the new denominator and write $\frac{5}{100} + \frac{20}{100}$ which equals $\frac{25}{100}$ or $\frac{1}{4}$.

KEY FACT   *To add fractions first rewrite them with identical denominators.*

The same method applies to subtracting fractions so that $\frac{1}{2}$ minus $\frac{1}{4} = \frac{2}{4} - \frac{1}{4} = \frac{1}{4}$.

Example

We will change the equation $\frac{1}{R} = \frac{1}{R_1} + \frac{1}{R_2}$ to get a formula for R.

The product $R_1 \times R_2$ can be used as the new denominator for the fractions $\frac{1}{R_1}$ and $\frac{1}{R_2}$ and we get $\frac{1}{R} = \frac{R_2}{R_1 \times R_2} + \frac{R_1}{R_1 \times R_2} = \frac{R_2 + R_1}{R_1 \times R_2}$

Then the reciprocal of $\frac{1}{R}$ ($= R$) must equal the reciprocal of the right-hand side of the equation so we turn the right side upside-down and get $R = \frac{R_1 \times R_2}{R_1 + R_2}$.

Test Yourself

1 Simplify these fractions (without use of decimals):

(a) $\frac{1}{2} + \frac{1}{4}$

> HINT
>
> *Use 4 for new denominator.*

(b) $\frac{1}{8} + \frac{1}{6}$

> HINT
>
> $\frac{1}{8} = \frac{3}{24}$

(c) $\frac{1}{10} - \frac{1}{20}$

> HINT
>
> $\frac{1}{10} = \frac{2}{20}$

(d) $\frac{1}{50} + \frac{1}{40}$

(e) $\frac{1}{10} - \frac{1}{200}$

2

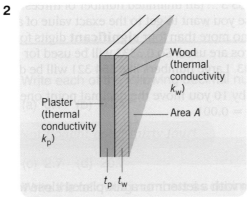

Plaster (thermal conductivity k_p)

Wood (thermal conductivity k_w)

Area A

t_p t_w

Fig. 7

Part of a wall consists of a sheet of wood coated with a layer of plaster (see *Fig. 7*). Calculate the thermal resistance (R) of unit area of this surface given the following information.

The units used are centimetres, milliwatts and kelvin degrees (equal to Celsius or centigrade degrees) and the formula needed is $R = \dfrac{t_w}{k_w \times A} + \dfrac{t_p}{k_p \times A}$.

$A = 1$, $t_w = 1$, $t_p = 1$, $k_w = 2$, $k_p = 10$. The unit for R will be kelvin per milliwatt.

> HINT
>
> $R = \frac{1}{2} + \frac{1}{10}$.

Exam Questions

Exam type questions to test understanding of Chapter 2

Exercise 2.5.2

1 A car travelling at steady speed covers 60 metres in 3 seconds. If the forces opposing its movement amount to 1400 newtons and the engine has an efficiency of 25% how much energy must be supplied to the engine per second.

HINTS AND TIPS

$$\text{Energy used per second for driving} = \text{power for driving} = \frac{work}{time}$$
$$= \frac{force \times distance}{time}.$$

Efficiency = energy per second successfully used (for driving) divided by total energy per second used (or supplied). Efficiency here is 25% or as a fraction $\frac{1}{4}$ so energy to be supplied per second is 4 times greater than is successfully used for driving.

2 (a) What is meant by the capacitance of a capacitor?
  (b) Calculate the capacitance of the combination of capacitors shown in *Fig. 8*.

4.7 µF   2.3 µF
(C_1)   (C_2)

Fig. 8

(AQA 1999, part question)

HINTS AND TIPS

$\frac{1}{C} = \frac{1}{C_1} + \frac{1}{C_2}$ *You can work with the data and units given. The answer for C will then be in µF.*

3 A surface is illuminated by a small lamp. The distance between the surface and the lamp is trebled so that the illumination of the surface becomes one ninth of its initial value. The lamp is now made 5 times brighter. What fraction is the resulting illumination of the original value?

4 Liquid emptying from a tank is reduced in depth to 7 tenths of its previous value in a minute. What fraction of the original depth is the final depth after (a) 2 minutes (b) 3 minutes?

HINT

For part (a): $\frac{7}{10}$ of original, then $\frac{7}{10}$ of this.

Answers to Test Yourself Questions

Exercise 2.1.1
1 (a) (i) $\frac{3}{4}$ kilogram (ii) $\frac{3}{8}$ litre (iii) $\frac{1}{3}$ litre
  (b) (i) $\frac{1}{3}$ (ii) $\frac{2}{3}$ ampere
2 (a) half (b) one sixth (c) a quarter litre
  (d) a hundredth
3 (a) 1/50 (b) 7/3 (c) 1/12 (d) 1/11
4 (a) 5 (b) 10 (c) 38

Exercise 2.2.1
1 (a) 2 (b) 3/4 (c) 46/7 (d) 43/60
2 (a) 1/2 (b) $\frac{4}{1}$ or 4 (c) 10 (d) $\frac{5}{1}$ or 5 (e) 1/5
  (f) 12 volt (g) 2/3 (h) 15

Exercise 2.2.2
1 (a) $\frac{2+4}{1}$ (b) $\frac{3+5}{4}$ (c) $\frac{8+3}{2}$ (d) $\frac{8+3}{2}$
  (e) $\frac{1}{2+2}$ (f) $\frac{3}{12+2}$ (g) $\frac{3}{2+1}$ (h) $\frac{6}{4+3}$

(i) $\frac{10}{1+1}$ (j) $\frac{60+15}{10+6}$ (k) $\frac{8-1}{4}$ (l) $\frac{5}{25-8}$
(m) $\frac{4x+7}{1}$ or $4x+7$ (n) $\frac{x-3}{3}$ (o) $\frac{5+3}{4}$

Exercise 2.3.1
1 (a) 5/3 (b) 10/7 (c) 11/9 (d) 14/5
2 (a) 1/12 of a volt (b) 2/25 (c) 23/100
3 (a) 1/8 (b) 1/120 (c) 2/15 (d) 12/35

Exercise 2.3.2
1 (a) $\frac{3}{4}$ (b) $\frac{7}{24}$ (c) $\frac{1}{20}$ (d) $\frac{9}{200}$ (e) $\frac{19}{200}$
2 $\frac{6}{10}$ or 0.6 K per mW

Exercise 2.4.1
1 (a) 6.8 (b) 11.11 (c) 1.22 (d) 9
2 (a) 6.001 (b) 2.205 (c) 282.5

Exercise 2.4.2

1 (a) Nought point one or one tenth
  (b) Nought point three four
  (c) Two point seven
  (d) Twenty point seven

2 (a) 0.01 (b) 0.3 (c) 53.2 (d) 205.08
  (e) 0.0115 (f) 0.23

3 (a) 0.1 (b) 0.02 (c) 0.034 (d) 0.5 (e) 0.1667

4 (a) $\frac{3}{10}$ (b) $\frac{7}{100}$ (c) $\frac{53}{100}$ (d) $\frac{8}{1000}$ or $\frac{2}{250}$ or $\frac{1}{125}$
  (e) $\frac{25}{1000}$ or $\frac{1}{40}$

Exercise 2.4.3

1 (a) 0.9 (b) 0.02 (c) 1.02 (d) 10.23 (e) 64.54

2 (a) 0.2 (b) 0.5

Exercise 2.5.1

1 (a) 25% (b) 2% (c) 3% (d) 120%

2 (a) 5% (b) 70% (c) 34.5%

3 (a) $\frac{20}{100}$ or $\frac{1}{5}$ (b) $\frac{10}{100}$ or $\frac{1}{10}$ (c) $\frac{150}{100}$ or $\frac{3}{2}$
  (d) $\frac{200}{100}$ or $\frac{2}{1}$ or 2

4 (a) 0.7 (b) 0.04 (c) 1.3

Chapter 3

Brackets and negative numbers

After completing this chapter you should:

- *understand why brackets are used, e.g. in the formula $P = \sigma A(T^4 - T_0^4)$*
- *have memorised rules for working with + and − signs*
- *have used your calculator with negative numbers*
- *understand the concept of directed numbers.*

3.1 Brackets

What are brackets used for?

If you were to write $2 \times 3 + 4$ it is likely to be interpreted as $6 + 4$ but you may have wanted the 3 and 4 to be added first to give 7 so that the 7 would be multiplied by 2 to give 14. We can use **brackets** to make this clear.

KEY FACT $2 \times (3 + 4)$ *means 2 times the enclosed* $3 + 4$ *expression.*

You leave out the $\times$ sign if no confusion will result and we write $2(3 + 4)$.
Keying this expression into your calculator and pressing the = button should give the answer as 14.

If you definitely want multiplication of the 2 and 3 to give 6 to be followed by adding the 4 you can used brackets to make this clear and you write $(2 \times 3) + 4$.

Here is an example of brackets being used in a physics calculation.

The formula for heat flow is

$$Q = \frac{kA(\theta_1 - \theta_2)}{L}$$

Suppose that values are obtained to be put in place of the letters, and these give $Q = 60 \times (20 - 5)$. Then Q is 60 times 15 and the answer is 900 joules.

A good rule is:

RULE *Work out the bracketed expression first.*

The term **bracket** is sometimes used for ( ) rather than the separate (and) symbols. Its meaning may also include the contained expression such as $(3 + 4)$.

When a multiplying sign is left out between two symbols or between a number and a symbol in expressions like $xy + 3$ or $4x + 9$ it is assumed that the product will be added to the other number and it is not necessary to put brackets around the product. (The x and y here represent numbers that will be entered when they are known. There are good examples of algebra in Chapter 8. The x and y values could be obtained from measurement of lengths or voltages, etc.)

It is not necessary to use brackets in an expression like $\frac{(x+2)}{3}$ because the dividing line is present beneath both the x and the 2, so both of these numbers must be divided by the 3. The brackets only indicate this same fact. However if the expression needs to be written all on one line then $(x + 2)/3$ is written to avoid confusion with $x + 2/3$ which would be 'x plus two thirds'.

In science we discuss the structure of atoms. Lithium atoms are quite simple and can help us learn more about brackets. *Fig. 1* shows a single lithium atom. In its nucleus it has 3 protons and 4 neutrons.

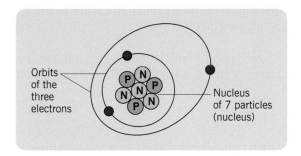

Orbits of the three electrons

Nucleus of 7 particles (nucleus)

Fig. 1 *A lithium atom.*

In two atoms there would be 2 times 3 for the number of protons and 2 times 4 for the number of neutrons. However we could also say that one atom contains $3 + 4$ particles in its nucleus and two such atoms contain $2 \times (3 + 4)$ or $2(3 + 4)$. So we have $2(3 + 4) = 2 \times 3$ plus 2×4.

This illustrates the rule:

RULE $a(b + c) = ab + ac$ where a, b and c can be any numbers.

Getting rid of the brackets in this way is often called 'multiplying out the bracket'. The importance of this multiplying-out rule will become more evident when you meet expressions like $2(x + 4)$, the x denoting an unknown number, and you want to remove the brackets. The rule gives $2(x + 4) = 2x + 8$.

Test Yourself   Exercise 3.1.1

Multiply out each of the following expressions.

1 (a) $2 \times (x + 5)$ (b) $7(x + 3)$ (c) $4(5 + x)$
2 (a) $3(2x + 5)$ (b) $22(2x + 4)$ (c) $4(50x + 10)$
3 (a) $2(x + 2 + 3)$ (b) $2(x + 2x + 5)$ (c) $9(x - 3)$

Separating out common factors

The number 6 is 2×3. The numbers which multiply to make the six are its **factors**. In the expression $ab + bc$ mentioned earlier the a is a factor of ab and it is a factor of ac. So a is a **common factor** of the two products. Now we saw that $a(b + c) = ab + ac$ and if we read it backwards we get $ab + ac = a(b + c)$ and the common factor has been separated out.

KEY FACT $ab + ac = a(b + c)$.

For another example consider the expression $6x + 4$ which could be part of a more complicated expression like $3x^2 + (6x + 4)$. As regards the bracketed expression 2 is a common factor of the 6 and 4 and therefore of the $6x$ and the 4. So all of the $6x + 4$ can be divided by 2 to give $3x + 2$. To compensate for this division the whole of the $3x + 2$, is multiplied by 2 to give $2(3x + 2)$.

$$6x + 4 = 2(3x + 2)$$

If the $2(3x + 2)$ is multiplied out in the way explained earlier in this chapter then $6x + 4$ is obtained and confirms the above equation.

The expression $7(3x + 6)$ is an example which already has a number multiplying its bracketed part but 3 is a common factor of the $3x + 6$ and can be separated out to give $7 \times 3 \times (x + 2) = 21(x + 2)$.

$$7(3x + 6) = 21(x + 2)$$

The formula $\alpha = \dfrac{R - R_0}{R_0 \theta}$ which concerns the effect of a temperature rise θ on the resistance of a conductor can be rearranged (see Chapter 8) to give $R = R_0 + R_0 \alpha \theta$. Now R_0 can be separated out to give $R = R_0(1 + \alpha\theta)$.

Test Yourself

Exercise 3.1.2

1 Rewrite each of the following expressions as a product:

HINT x^2 means $x \times x$

(a) $5x + 10$ (b) $2x - 2$ (c) $3x + x^2$ (d) $36x^2 + 6x$
(e) $IR + Ir$ (= EMF of a voltage supply with internal resistance r and connected to a resistance R)
(f) $I^2R + I^2r$ (= electrical power supplied)

Products of brackets

Sometimes when a formula is being obtained a product is met that has the form $(a + b)(c + d)$ where a, b, c and d are any numbers that will be supplied when the formula is used. We need then to be able to rewrite the expression without brackets – we need to multiply out the brackets.

The rule is:

RULE $(a + b)(c + d) = ac + ad + bc + bd$.

The lines that have been drawn between the symbols may help you to see which letters are multiplied. Letter a is multiplied first by the c in the other bracket then by d. The b does the same.

Often a product of brackets contains only two unknown quantities and $(x + 3)(y + 4)$ is an example. As will be explained in Chapter 8, x and y are favourite symbols when there is no special reason for using different letters although a, b, c, d may be preferred when more symbols are needed. For this example

$$(x + 3)(y + 4) = xy + 4x + 3y + 12$$

For $(x + 5)(x + 4)$ there is only one unknown. Removing the brackets gives

$$(x + 5)(x + 4) = xx + 4x + 5x + 20$$

As will be explained in Chapter 4 the x times x is written as x^2 and if we add the $4x$ and $5x$ to get $9x$ we have

KEY FACT $(x + 5)(x + 4) = x^2 + 9x + 20$.

Another fact that you may have noticed in the statements above is that a number represented by a symbol and multiplied by a number is written with the number first. So we write $4x$ not $x4$.

Test Yourself   Exercise 3.1.3

Multiply out each expression:

1 (a) $(x + 5)(y + 3)$ (b) $(x + 1)(y + 1)$ (c) $(2x + 7)(3y + 2)$
2 (a) $(\frac{x}{2} + 3)(y + 2)$ (b) $(\frac{x}{3} + 4)(\frac{y}{3} + 4)$ (c) $(x + 3)(\frac{1}{x} + 2)$

3.2 Plus, minus and directed numbers

In a previous calculation of heat flow we had $Q = \dfrac{kA(\theta_1 - \theta_2)}{L}$. The temperature at which water normally freezes is zero degrees on the Celsius scale (0 °C). A lower temperature than this is a minus temperature or negative temperature. Suppose that you are told that $\theta_2 = -5$. How do you deal with the $(\theta_1 - -5)$ that you get? We can answer this question when **directed numbers** have been explained. Directed numbers are also important for indicating the directions of vector quantities which are quantities such as forces that have directions as well as sizes.

In this book until now a + (positive) sign has been used to denote *plus* or *add* while a − (negative) sign denoted *minus* or *take away* or *subtraction*. The result of adding is called the **sum** and the result of subtracting is the **difference**. You may also have regarded + as *above* and − as *below*, (for example when describing temperatures above and below 0 °C). But now we can extend the use of the + and − signs to indicate direction. The numbers affected by the signs are then called **directed numbers**.

Direction is important when dealing with quantities that can exactly either oppose each other or assist each other, such as two pushes (forces) acting on an object. They can be oppositely directed and if of equal size can even cancel in their effects.

There are various ways in which direction can be introduced and the following explanation is just one suggestion. Please read this explanation but note that it is the resulting rules that are important and must be remembered.

Think first about heights. Imagine that you are to climb a ladder and descend as instructed by a mathematical expression. You will begin at zero which we could decide is ground level. An initial direction must also be decided. Let this be *up*.

+ and – signs will be used as follows:
– will mean 'reverse your direction' and + will mean 'do not change direction'

Consider +5.

For any expression you start at zero and face the initial direction. The + before the 5 means 'don't change direction', the 5 means move 5 units, e.g. 5 metres. The result is that you reach a height of 5 up from zero.
After each move you have to face the initial direction, turning if necessary.

Now consider –3.

Start at zero. Face the initial direction (up). The – means 'reverse your direction' so you are now facing downwards. Move 3. Result is 3 down from zero. Now turn to face the initial direction.
These results do not conflict with the idea of + 5 meaning 5 above and – 3 meaning 3 below.

Since +5 produces a movement in the chosen initial direction this direction is usually called the **positive direction** and the +5 is described as a positive number.

KEY FACT *When no sign is shown before a number a + sign is assumed so 5 means +5.*

Expressions like 5 + 5 have been met previously (as '5 plus 5'). The same expression can describe a directed procedure.

For the +5 + 5 the first +5 results in a height of 5 up from zero as explained above and leaves you facing upwards. The second + sign says 'no change of direction' and then the second 5 takes you to a final height of 10 up from zero. The result is the same as for +10. The +5 + 5 = + 10 or 5 + 5 = 10 illustrates the agreement between the use of + and – signs for directions and their use for adding and subtraction.

For 5 – 3 the 5 takes you to a height of 5 above zero and facing up. The – tells you to reverse direction (now facing down). The 3 moves you to a final height of 2 up from zero. So 5 – 3 = +2 or just 2. This is illustrated in *Fig. 2(a)*.

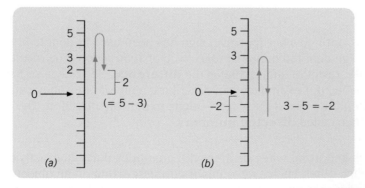

Fig. 2 Directed numbers (a) 5 – 3 = +2 (b) 3 – 5 = –2

Using the same principles $3 - 5 = -2$ (see *Fig. 2(b)*) and $-5 + 3 = -2$ and $-5 - 3 = -8$, all these results being as expected.

Directed numbers introduce new ideas when an expression like $5 + x$ is met where x denotes a number to be specified. If this number is then given as -2 we need to know how to deal with $+ -2$.

From what has been said $+-$ will be the same as $-$ because the $+$ causes no change but the following $-$ sign makes a reversal. So $5 + -2$ is 3. No change of direction is required for $+ +$ and it is the same as a single $+$. For $-+$ the meaning is the same as a single $-$ and $- -$ has the same effect as a $+$ because a reversal followed by a reversal restores the original direction.

The rules for pairs of signs are $++$ means $+$, $- -$ means $+$, $+ -$ means $-$, $- +$ means $-$. These rules should be memorised.

> **RULES** *Two different signs together mean $-$ and two signs the same mean $+$.*

Are directed numbers important in physics?

Well, forces can be up and down or directed to left and right or be zero for example. Velocities will have signs and their directions are very important in collisions and temperatures can be above or below zero.

Now we can see what to do about the $(\theta_1 - -5)$ mentioned earlier. The $- -$ means $+$ so $(\theta_1 - -5) = (\theta_1 + 5)$.

It is usually obvious or has been clearly stated which side of zero is the positive direction. For heights the positive direction is usually *up* as suggested already. For temperatures it is always the warmer side of zero.

Test Yourself

Exercise 3.2.1

Simplify each of the following expressions:

1 (a) $8 + +5$ (b) $3 - -2$ (c) $6 + -2$ (d) $7 - +3$ (e) $5 + +8$

2 (a) $5 + -8$ (b) $2 - -3$ (c) $2 + -6$ (d) $3 - -7$

3 (a) $2x + x$ (b) $2x - +x$ (c) $2x - x$ (d) $4x - +5x$ (e) $5x - +4x$

    (f) $5x - -x$ (g) $3 + -5 - -2$ (h) $-1 - 1 - 1$ (i) $9x - +3x + +5$

Multiplying with negative numbers

When discussing directed numbers -3 was a measurement in the opposite direction to 3 or $+3$. It is also true that -3, which is $-1 - 1 - 1$ is 3 times -1, or $-3 = -1 \times 3$.

> **KEY FACT** *Multiplying a number by -1 reverses its direction.*

It follows that $3 \times -1 \times -1 = 3$ because the two reversals restore the original direction and $-5 \times -3 = -1 \times 5$ times -1×3 which equals 15. This result shows that a negative number multiplied by a negative number gives a positive product. This rule and rules for $+$ times $+$ and for $+$ times $-$ are similar to the rules for adding and subtracting with $+$ and $-$ signs.

These rules are:

- A + number multiplied by a + number results in a + product.
- A + number multiplied by a − number results in a − product.
- A − number multiplied by a + number results in a − product.
- A − number multiplied by a − number results in a + product.

> **RULES** *Multiplying the same signs produces a single + sign. Multiplying different signs produces a single − sign.*

These rules will be particularly useful when − signs appear in bracketed expressions as the following example illustrates. The above rules for multiplying signs are illustrated by this example if the brackets are multiplied out. You get

$$(5-3)(7-4) = +35 - 20 - 21 + 12$$

The 5 and 7 are read as +5 and +7. So +5 × +7 becomes +35, the +5 × −4 involves two different signs and gives a − product, the −3 and +7 multiply to a − product and the −3 × −4 involves *same* signs so that the 12 is positive.

The procedure is seen to work because the $35 - 20 - 21 + 12$ equals 6 and you can see that this is the correct answer because the first bracket equals 2, the second bracket equals 3 and the product is 6.

What about dividing numbers with signs?

A negative denominator can be made positive by multiplying top and bottom of the fraction by −1. This has the effect of putting a multiplying −1 into the numerator. So for example:

$$\frac{5}{-3} = \frac{5 \times -1}{-3 \times -1} = \frac{5 \times -1}{3} = \frac{-5}{3}$$

> **KEY FACT** *A − sign multiplying a denominator can be moved so that it multiplies the numerator instead.*

Test Yourself

Exercise 3.2.2

1 Simplify each of the following products.

   (a) 4×5 (b) $+4 \times +5$ (c) 4×-5 (d) 5×-4
   (e) -4×-5 (f) $-4 \times -5 \times -6$ (g) $-4 \times 5 \times -6$

2 Simplify:

   (a) $\frac{8}{4}$ (b) $\frac{+8}{+4}$ (c) $\frac{-8}{2}$ (d) $\frac{8}{-2}$
   (e) $\frac{-8}{-2}$ (f) $-8 \times \frac{-3}{2}$ (g) $-8 \times \frac{-3}{-2}$ (h) $\frac{+8}{-2}$

Test Yourself

Exercise 3.2.3

Multiply out:

1 (a) $(x+4)(y-2)$ (b) $(x-5)(y+4)$ (c) $(x-6)(y-3)$
2 (a) $(2x-3)(3y-2)$ (b) $(\frac{1}{5}-5)(\frac{2x}{3}+2)$ (c) $2(x+1)(y-1)$

Entering a negative number into a calculator

On the *fx-83WA* calculator a negative number can be obtained by pressing the subtraction key (marked −) before the number. So −2 would use − followed by 2 and −2 − 3 would use the same subtraction key twice and give the correct answer of −5.

An alternative is $\boxed{-}\boxed{2}\boxed{+}\boxed{(-)}\boxed{3}\boxed{=}$

The (−) makes a number negative.

KEY FACT   *To enter a negative number use the (−) key or −.*

For practice with these keys and for revision of the topics covered so far in this chapter you could try additions, subtractions, multiplications and divisions involving negative numbers. If you use simple numbers you will know the answers to expect from your calculator. $\frac{-10}{5}$, and −2 + 3, and −2 × −3 could be tried first.

A further alternative for a multiplication like −2.1 × −3.2 is to use the calculator without the minus signs and then insert the appropriate sign into the answer. For 2.1 × 3.2 the product is 6.72 and the product of two minus signs is +. Hence the answer is +6.72 or just 6.72. For −2.1 − 3.2 there are two numbers 2.1 and 3.2 both in the same direction and amount to 5.3 in the negative direction. The answer is −5.3. This could be worked out as 2.1 + 3.2 on the calculator to get 5.3 and the minus sign is added afterwards.

$$-2.1 - 3.2 = -1(2.1 + 3.2)$$

Negative numbers will be very important in Chapter 4.

A special result, $(a + b)(a - b) = a^2 - b^2$

Applying the rules just described to a product $(x + 3)(x - 3)$, i.e. to an expression of the form $(a + b)(a - b)$ where a and b are any two numbers, we get

$$(x + 3)(x - 3) = xx + 3x - 3x - 9$$

and the $+3x - 3x$ equals 0 so that the product becomes $xx - 9$. We can use the fact that x times x is written as x^2 and is called **x squared**. So the result obtained is $x^2 - 9$.

In general, this means, not considering any particular numbers but using the symbols a and b for the numbers the special result is:

KEY FACT $(a + b)(a - b) = a^2 - b^2$.

It is special because the product has only two terms whereas similar products have more. The product $(a + b)(a + b)$ for example equals $a^2 + ab + ba + b^2$ which, although it simplifies to $a^2 + 2ab + b^2$, still has more than two terms.

3.3 Some special uses of + and − signs and brackets in physics

Using products of brackets in physics

In the theory of interference of light, when the light passes through two adjacent slits as shown in *Fig. 3*, the difference between distances AC and BC is important.

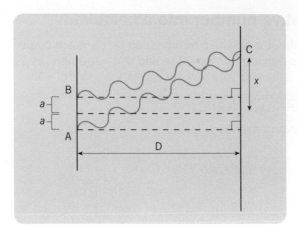

Fig. 3 *Diagram for two-source interference.*

From *Fig. 3* it can be deduced by using the Pythagoras formula which will be met on page 141 that

$$BC^2 = (x - a)^2 + D^2 \qquad \text{and} \qquad AC^2 = (x + a)^2 + D^2$$

So $AC^2 - BC^2 = (x + a)^2 - (x - a)^2$ But $AC^2 - BC^2 = (AC - BC)(AC + BC)$ because of our $(a + b)(a - b) = a^2 - b^2$ rule and using our multiplying-out rules shows that $(x + a)^2 = x^2 + 2ax + a^2$ and $(x - a)^2 = x^2 - 2ax + a^2$ so that

$$(AC - BC)(AC + BC) = AC^2 - BC^2 = (x + a)^2 - (x - a)^2 = (x^2 + 2ax + a^2) - (x^2 - 2ax + a^2)$$
$$= 2ax - - 2ax = 4ax$$

In practice A and B may be very close together and cause AC and BC both to almost equal D so that their sum is close to 2D. So $2D(AC - BC) = 4ax$ and we have a relationship between the path difference $AC - BC$ and the distance x.

A calculation concerning a collision between two objects showed that the velocity (v_1) of one of the objects after collision is given by the equation $11v_1^2 - 2000v_1 + 90\,000 = 0$. This equation can be rewritten as $(11v_1 - 900)(v_1 - 100) = 0$ For this to be true one of the bracketed expressions must be zero or both. Consequently $v_1 = 100$ or $11v_1 = 900$ so that $v_1 = 81.8$ metre per second.

Exam Questions

Exam type questions to test understanding of Chapter 3

Exercise 3.3.1

1 The resistance of a coil of wire increases with temperature according to the formula $R = R_0(1 + \alpha\theta)$. R_0 is the resistance at 0 °C, θ is the temperature for which the resistance is R and α is the *temperature coefficient of resistance* for the coil concerned.

Calculate R for $R_0 = 4.0$ ohms, $\theta = 80$ degrees and $\alpha = \frac{5}{1000}$ per degree.

2 Calculate Q given that $Q = 60(\theta_1 - \theta_2)$ and $\theta_1 = 20$, $\theta_2 = -5$. Q will be in watts.

3 A light rod 2.0 m long rests horizontal on a pivot with a 3.0 newton force acting downwards on it at one end and a similar force of 2.0 newton at the other end. At what distance (x) from the 2.0 newton force must the pivot be placed?

HINT   *The equation needed is $2x = 3(2 - x)$.*

**Answers to
Test Yourself
Questions**

Exercise 3.1.1
1 (a) $2x + 10$ (b) $7x + 21$ (c) $20 + 4x$
2 (a) $6x + 15$ (b) $44x + 88$ (c) $200x + 40$
3 (a) $2x + 4 + 6$ or $2x + 10$
   (b) $2x + 4x + 10$ or $6x + 10$ (c) $9x - 27$

Exercise 3.1.2
1 (a) $5(x + 2)$ (b) $2(x - 1)$ (c) $x(3 + x)$ (d) $6x(6x + 1)$
   (e) $I(R + r)$ (f) $I^2(R + r)$

Exercise 3.1.3
1 (a) $xy + 3x + 5y + 15$ (b) $xy + x + y + 1$
   (c) $6xy + 4x + 21y + 14$
2 (a) $\frac{xy}{2} + x + 3y + 6$ (b) $\frac{xy}{9} + \frac{4x}{3} + \frac{4y}{3} + 16$
   (c) $1 + 2x + \frac{3}{x} + 6$ or $2x + \frac{3}{x} + 7$

Exercise 3.2.1
1 (a) 13 (b) 5 (c) 4 (d) 4 (e) 13

2 (a) 13 (b) 5 (c) -4 (d) 10
3 (a) $3x$ (b) x (c) x (d) $-x$ (e) x (f) $6x$
   (g) 0 (h) -3 (i) $6x + 5$

Exercise 3.2.2
1 (a) 20 (b) 20 (c) -20 (d) -20
   (e) $+20$ or just 20 (f) -120 (g) 120
2 (a) 2 (b) 2 (c) -4 (d) -4 (e) 4
   (f) 12 (g) -12 (h) -4

Exercise 3.2.3
1 (a) $xy - 2x + 4y - 8$ (b) $xy + 4x - 5y - 20$
   (c) $xy - 3x - 6y + 18$
2 (a) $6xy - 4x - 9y + 6$
   (b) $\frac{2x}{15} + \frac{2}{5} - \frac{10x}{3} - 10$ or $-3.2x - 9.6$
   (c) $2xy - 2x + 2y - 2$

On a calculator a number like 150 000 000 000 (the radius of the Earth's orbit in metres) is too long for the screen to display but can be shown as $150 \times 10^9 = 1.5 \times 10^{11}$. This way of writing numbers is used in a calculator's *scientific mode* mentioned in Chapter 1 and is mentioned again later in this chapter. The *fx-83WA* calculator simplifies the 1.5×10^{11} to 1.5^{11}.

To change your *fx-83WA* calculator to scientific mode use the mode key as explained in Chapter 1 and select the Sci mode. When you do this you will immediately be presented with a choice of numbers from 0 to 9. Choose 4. Answers will then be given as a four figure number followed by a power of ten. (The 4 selected is in fact the number of significant figures required for your answers, as explained in Chapter 5.)

KEY FACT   *You should, when working with all further chapters of this book, employ the **scientific mode** of your calculator.*

Test Yourself

Exercise 4.1.1

1 Write as a power of ten

(a) 100   (b) 1 million   (c) 10 000   (d) 100 000

2 Write as an ordinary decimal number (not a power)

(a) 10^3 (b) 10^5 (c) ten squared   (d) 10 cubed

3 Express as a simple decimal number

(a) 3^2 (b) 2^3 (c) 4^2 (d) 2^5

HINT *$2 \times 2 \times 2 \times 2 \times 2$*

4 Write as a single power

(a) $10^2 \times 10^4$

HINT *10×10 times $10 \times 10 \times 10 \times 10$*

(b) $2^2 \times 2^3$ (c) $1000 \times 100 \times 1000$ (d) $x \times x^3$

5 Write as an ordinary number (not a power)

(a) 7×10^3 (b) 7.4×10^4 (c) 33.2×10^4 (d) 2.13×10^2

10 000 or 10^4 kilograms

Fig. 2 Powers of ten in use.

Working with powers

If you meet a product of powers that have the *same base*, e.g. $10^2 \times 10^3$ the product can be written as a single power with the same base and an exponent that is the sum of the original exponents, i.e. 10^5.

$$10^2 \times 10^3 = 10 \times 10 \text{ times } 10 \times 10 \times 10$$

or, in the general case, using *a* and *b* to represent the two exponents

RULE $10^a \times 10^b = 10^{a+b}.$

Here is a typical example of this rule being used:

$$\text{Mass of water} = \text{volume} \times \text{density}$$
$$= 5.0 \times 10^2 \text{ m}^3 \times 1.000 \times 10^3 \text{ kg per m}^3$$
$$= 5.0 \times 10^5 \text{ kg}$$

It is this rule that shows $10^0 = 1$ by letting $b = 0$.

Here is another rule:

RULE $(a \times b)^c = a^c \times b^c.$

You might, for example, see $(3.2 \times 10^5)^2$ changed to $3.2^2 \times (10^5)^2$. The rule can be understood best by a simpler example, like $(3 \times 4)^2$. This equals $12^2 = 144$ of course but we can write

$$(3 \times 4)^2 = 3 \times 4 \times 3 \times 4 = 3 \times 3 \times 4 \times 4 = 3^2 \times 4^2.$$

In Chapter 5 this rule is used for some *dimensions* calculations.

If we have an expression of the form $(a^b)^c$, for instance $(2^3)^2$ then it can be rewritten a^{bc} and $(2^3)^2$ equals 2^6. This agrees with $(2^3)^2 = 2 \times 2 \times 2$ times $2 \times 2 \times 2$ which means 2^6.

KEY FACT $(a^b)^c = a^{bc}.$

Now to see how this fact may be useful.

The Earth is approximately a sphere with a radius of 6.4×10^3 kilometres and its mean density is 5.5×10^3 kilograms per cubic metre. What approximately is its mass?

Here we need first to calculate the volume using the formula $V = \dfrac{4}{3} \pi \times \text{radius}^3$ and the radius is 6.4×10^3 km $= 6.4 \times 10^3 \times 1000$ metres.

$$\text{So } V = \frac{4}{3} \times \pi \times (6.4 \times 10^6)^3 = 4.189 \times 6.4^3 \times (10^6)^3 = 4.189 \times 262.1 \times 10^{18}$$
$$= 1.098 \times 10^{21} \text{ cubic metres}$$

The mass = volume × density
$$= 1.098 \times 10^{21} \times 5.5 \times 10^3 = 6039 \times 10^{21} \text{ or } 6.04 \times 10^{24} \text{ kilogram}$$

Test Yourself

Write each of the following products as a single power:

1 (a) $10^3 \times 10^4$ (b) $10^3 \times 10^3$ (c) $10^5 \times 100$ (d) $10^2 \times 10^2 \times 10^8$ (e) $2^4 \times 2^6$

2 Complete each of the following equations:

(a) $(4 \times 2)^2 = 4^2 \times \ldots$ (b) $(a \times b)^2 = a^2 \times \ldots$ (c) $(p \times 7)^2 = \ldots \times 49$
(d) $(7p)^2 = \ldots \times p^2$ (e) $(4b)^2 = \ldots \, b^2$ (f) $(4 \times 10^8)^2 = \ldots \times 10^{16}$

3 Simplify the following expressions:

(a) $(2^3)^2$ (b) $(2 \times 10^4)^3$ (c) $(a^{2b})^4$ (d) $(p^q)^2$

Negative powers

When dimensions are discussed in Chapter 5 it will be necessary to write $\frac{1}{L^2}$ as L^{-2} for example. Much more important are negative powers of 10 because they enable us to work easily with very small numbers. A tenth (1/10) is written as 10^{-1}. Why is this?

Well we get a clue from $\frac{1}{10} \times 1000 = 100$.

For this to agree with the rule regarding adding of exponents (**indices**) when powers are multiplying, the 1/10 must equal 10^{-1} so that $10^{-1} \times 10^3 = 10^2$, the indices − 1 and 3 adding to give the 2. So, as mentioned earlier,

KEY FACT $\frac{1}{10} = 10^{-1}$.

Similarly for $\frac{1}{100} \left(= \frac{1}{10 \times 10} \right)$ the power is −2, for $\frac{1}{1000} \left(= \frac{1}{10 \times 10 \times 10} \right)$ it is −3. The exponent equals the number of tens in the denominator of the fraction and the − sign can be regarded as indicating 'in the denominator'.

A useful observation here is that

$$\frac{1}{10^{-2}} = \frac{1}{1/100} = 100 = 10^{+2} \text{ or } 10^2$$

So the **reciprocal** of 10^{-2} is 10^2 and in general

KEY FACT   *The reciprocal of 10^x is 10^{-x}.*

0.001 is a decimal fraction and from what has been said above it equals 10^{-3}. Now a number like 0.002 can be written as 2×0.001 or 2×10^{-3}. In the same way 0.02 equals 2×10^{-2}.

When writing these **decimal fractions** to involve powers the negative exponent equals the number of places that the decimal point is moved to the right.

KEY FACT   *$0.0021 = 0002.1 \times 10^{-3}$ meaning 2.1×10^{-3}. The 3 equals the number of places the decimal point has moved to the right.*

The rule explained earlier for products of powers still applies when one or more of the powers concerned is negative. Negative powers of ten are very useful in physics for small quantities such as the electric charge on an electron which is 1.6×10^{-19} coulombs in size. So for example $10^5 \times 10^{-2} = 10^3$ because the exponents (5 and −2) are added for a product of powers.

Suppose that you want to divide two powers that have the same base. Probably the base will be 10 of course. An example is $\frac{10^5}{10^2} \left(= \frac{10 \times 10 \times 10 \times 10 \times 10}{10 \times 10} \right)$. This expression simplifies to $10 \times 10 \times 10 = 10^3$. The index for the denominator (2) is subtracted from the numerator's index (5) to find the index for the answer.

RULE $\frac{10^a}{10^b} = 10^{a-b}$.

This rule also agrees with the $10^a \times 10^b = 10^{a+b}$ rule as can be seen from

$$\frac{10^a}{10^b} = 10^a \times \frac{1}{10^b} = 10^a \times 10^{-b} = 10^{a+-b} = 10^{a-b}.$$

Some examples of this are:

$$\frac{10^6}{10^2} = 10^{6-2} = 10^4$$

$$\frac{10^{-3}}{10^2} = 10^{-3-2} = 10^{-5}$$

$$\frac{10^{-7}}{10^{-4}} = 10^{-7--4} = 10^{-7+4} = 10^{-3}$$

$$\frac{10^3}{10^{-2}} = 10^{3--2} = 10^{3+2} = 10^5$$

$$\frac{4.1 \times 10^{-2}}{10^3} = 4.1 \times \frac{10^{-2}}{10^3} = 4.1 \times 10^{-2-3} = 4.1 \times 10^{-5}$$

A consequence of using the $a^b \times a^c = a^{b+c}$ rule for a calculation like 2×10^3 multiplied by 4×10^{-3} is that you get an answer (8×10^0) that involves 10^0. This may at first seem strange.

However, when you think of the calculation as $2000 \times \dfrac{4}{1000}$ you know it equals 8. So $10^0 = 1$.

This idea is found to fit in with other powers if a list of powers of ten is made beginning with $1000 = 10^3$ then $100 = 10^2$ and $10 = 10^1$ and to continue the sequence $1 = 10^0$.

Note again the key fact

KEY FACT $10^0 = 1$ and $10^1 = 10$.

Test Yourself

1 Write as powers of ten:

(a) $\frac{1}{10}$ (b) $\frac{1}{10\,000}$ (c) $1/10^{-4}$

2 Write as a single power:

(a) $\frac{1}{10} \times 10^3$ (b) 1000×10^{-2} (c) $10^4/10$

3 What is the reciprocal of:

(a) 10^{-4} (b) 10^2 (c) 10^{-19}

4 Write as a single power:

(a) $10^{-8} \times 10^{12}$ (b) $10^{-3} \times 10^{-4}$ (c) $x^{-2} \times x^{-2}$

> **HINT**   *Add indices.*

5 Simplify each of the following expressions:

(a) $\dfrac{20 \times 10^5}{2 \times 10^4}$

> **HINT**   *Cancelling gives* $10 \times \dfrac{10^5}{10^4}$

(b) $\dfrac{45 \times 10^{-4}}{5 \times 10^2}$ (c) $\dfrac{42 \times 10^{-6}}{6 \times 10^{-4}}$ (d) $\dfrac{2.6 \times 10^3}{1.3 \times 10^5}$

Example

A hosepipe which has a cross-section area of 4.0×10^{-4} m^2 is connected to a pump and is producing a jet of water travelling horizontally at 5.0 m s^{-1}. The density of water $= 1.0 \times 10^3$ kg m^{-3}. Calculate the mass of water leaving the pipe per second.

Answer

(Units such as **m s^{-1}** and **kg m^{-3}** are explained in Chapter 5.)
The volume per second leaving the pipe

= area × length of water leaving per second
$= 4 \times 10^{-4} \times 5 = 20 \times 10^{-4}$ cubic metre per second.

The mass *m* of water leaving per second

= volume × density $= 20 \times 10^{-4} \times 1.0 \times 10^3 = 20 \times 10^{-1}$ (adding the exponents, $-4 + 3 = -1$).

The mass is therefore 2 kilograms.

Adding and subtracting numbers that include powers of ten

It will often be useful to add two numbers like 7.0×10^3 and 2.0×10^4 without using a calculator. It may for example be quicker to do this. The procedure required is to first rewrite the numbers with the same power of ten in both numbers.

The $2.0 \times 10^4 = 20 \times 10^3$ (20 thousand) and adding this to the 7.0×10^3 (7 thousand) gives 27×10^3 (27 thousand).

> **KEY FACT** *7.0×10^3 plus $2.0 \times 10^4 = (7.0 \times 10^3) + (20 \times 10^3)$*
> *$= 27 \times 10^3$.*

(As will be explained in Chapter 7, this answer is best written as 2.7×10^4.) It would be quite wrong in this last calculation to add the 7.0 and 2.0 to get 9.0 as part of the answer.

> **KEY FACT** *For adding or subtracting numbers that include powers of ten first make the powers the same.*

Test Yourself

Exercise 4.1.4

1 Convert each of the following expressions to a single product.

(a) $(2.1 \times 10^7) + (190 \times 10^5)$
(b) $(3 \times 10^{-4}) + (5 \times 10^{-5})$
(c) $(8 \times 10^5) - (8 \times 10^4)$
(d) $(0.050 \times 10^{-5}) + (0.003 \times 10^{-5})$
(e) $(0.80 \times 10^{-4}) - (0.03 \times 10^{-5})$

2 Show that $\dfrac{3.6 \times 10^{-5}}{0.12 \times 10^3} + (0.40 \times 10^{-6}) = 70 \times 10^{-8}$

Example

Simplify the expression $\dfrac{5.7 \times 10^{-7} \times 3.3 \times 10^5}{2.9 \times 10^{-2}}$.

Answer

$$\frac{5.7 \times 10^{-7} \times 3.3 \times 10^5}{2.9 \times 10^{-2}} = \frac{5.7 \times 3.3}{2.9} \times \frac{10^{-7} \times 10^5}{10^{-2}}$$

$$= 6.5 \times 10^{-7+5--2}$$

$$= 6.5 \times 10^0 = 6.5$$

In physics exponents are often whole numbers and can then be handled without calculators.

Powers of ten in tables of experiment results

In a table of experimental results it is tedious and even unsightly to repeat the units with every result shown in the table. It is better to write the unit concerned in the heading of each column. As will be seen in Chapter 5, a quantity divided by its unit is a pure number, and has no units. So if the heading says that each result in its column is the quantity divided by its unit then units need not be shown against each result. *Fig.* 3a shows headings of a table conforming to these suggestions. The quantities listed are force F and acceleration a. The unit for a is written as m s^{-2} and means **metres per second per second**. The values of the force F are small but can be expressed in millinewtons (mN) each being a thousandth of a newton. Powers of ten are then not involved.

F/mN	a/m s^{-2}
15	0.60
20	0.81
25	0.99

(a)

$\dfrac{F \times 10^3}{N}$	$\dfrac{a \times 10^{-25}}{m\ s^{-2}}$
0.20	22
0.40	45
0.60	61

(b)

Fig. 3 *Headings for tables of results. (a) A table of experiment results. (b) Results for a different experiment.*

In *Fig.* 3b powers of ten are needed because a is an enormous number of m s^{-2}. If each value of a is multiplied by 10^{-25} the resulting sizes are a convenient number of m s^{-2} and it is these values that have been recorded in the table.

The forces here are, as before, of mN sizes and could be recorded in mN but the choice has been made to use powers of ten and record the force $\times 10^3$ which amounts to 0.20 newton, etc.

The displaying of quantity divided by unit fits in well with the use of powers of ten. Where $F \times 10^3$ was recorded in newtons (N) the same figures would be recorded if F itself were recorded using for its unit the millinewton which is $\times 10^{-3}$ N. So the table heading could be $\dfrac{F \times 10^3}{N}$ or $\dfrac{F}{N \times 10^{-3}}$. These headings mean the same and are the same according to the rules of mathematics because 10^3 multiplying in a numerator is the same as 10^{-3} multiplying in the denominator.

Displaying results in this way has for a long time been the preferred method for A level examination papers and hopefully for all physics authors. This way of handling measurements is discussed again when the labelling of graphs is explained.

Using a calculator for powers of ten

To enter a number with an exponent such as 10^5 the key marked 10^x can be used. The 10^x on the recommended calculator is the second function of the key used for **log** and requires the

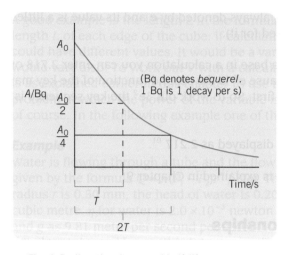

(Bq denotes *bequerel*.
1 Bq is 1 decay per s)

Fig. 4 *Radioactive decay and half-life.*

Answer

$A = A_0 e^{-\lambda t}$. $A = 30\ 000$, $\lambda = 0.20 \times 10^{-3}$ and $t = 5$ minute $= 5 \times 60$ second $= 300$ s.

$$\therefore \quad \lambda \times t = 0.20 \times 10^{-3} \times 300 = 0.060$$

$$\therefore \quad A_0 = \frac{A}{e^{-\lambda t}} = \frac{30\ 000}{e^{-0.06}} = \frac{30\ 000}{0.9418} = 31\ 855 \text{ or approximately 32 thousand per second.}$$

Test Yourself

Exercise 4.3.1

1 The half-life (T) of a strontium 90 beta radiation source is 20 years. If the initial activity of a certain source is 40 000 disintegrations per second what will its activity (A) be after a time (t) of   (a) 20 years   (b) 40 years   (c) 10 years?

HINT $A = A_0 2^{-t/T}$

2 A beta radiation source has a disintegration constant or decay constant λ of 0.18 per minute. What fraction (A/A_0) of the initial activity will remain after a time of 10 minutes? You may wish to use the formula $A = A_0 e^{-\lambda t}$.

HINT *For e see above text.*

4.4 Standard form

Standard form and modes on your calculator

Standard form is a method of presenting numbers. It requires that a number like 3759 should be written as 3.759×10^3 because the rules for standard form are:

KEY FACT *The number is written with **one digit** in front of the decimal point followed by a multiplying **power of 10** as appropriate.*

Here are some examples.

$375\,900 = 3.759 \times 10^5$ $0.000\,375\,9 = 3.759 \times 10^{-4}$

$3.759 = 3.759$ (because $10^0 = 1$ and does not have to be displayed)

Standard form has a special advantage for showing the accuracy of a measurement and this will be explained in Chapter 7. Standard form is used for displaying answers when calculators are working in *scientific mode*.

KEY FACT   *Scientific mode* displays answers in **standard form**. *For all future chapters use of scientific mode is assumed.*

Test Yourself   Exercise 4.4.1

Write each of the following numbers using standard form.

(a) 5555   (b) 0.005 555   (c) 0.8354   (d) 954 million

(e) $\frac{1}{23}$

HINT   *Use calculator.*

(f) $\frac{1}{354}$

HINT   *Use calculator.*

(g) A quarter of a million.

HINT   250 000

Exam Questions   Exam type questions to test understanding of Chapter 4

Exercise 4.4.2

1 A mass of 0.80 kg suspended from a vertical spring oscillates with a period of 1.5 s. Calculate the force constant of the spring.

(Edexcel 2000, part question)

HINT   *Period $T = 2\pi\sqrt{\dfrac{m}{k}}$ so that $k = 4\pi^2 m/T^2$. You want k. Unit for k will be N m^{-1}.*

2 A radioactive source produces 10^6 α-particles per second. When all the ions produced in air by these α-particles are collected, the ionisation current is about 0.01 μA. If the charge on an ion is about 10^{-19}, what is the best estimate of the average number of ions produced by each particle?

A 10^5 **B** 10^6 **C** 10^7 **D** 10^8

(OCR 2000)

HINT   *You need to work out $\dfrac{0.01 \times 10^{-6}}{10^6 \times 10^{-19}}$.*

3 One of the lines in the sodium spectrum has wavelength 5.9×10^{-7} m. Given that velocity of light = 3.0×10^8 m s^{-1} and Planck constant = 6.6×10^{-34} J s. Which one of the A to D below is the energy, in J, of a photon, wavelength 5.9×10^{-7} m?

A 2.7×10^{35} **B** 3.4×10^{-19} **C** 1.2×10^{-31} **D** 3.7×10^{-36}

(OCR 2000)

> HINT
>
> *Photon energy = $\frac{hc}{\lambda}$ where h is the Planck constant, c the velocity of light and λ is the wavelength.*

4 The initial activity (A_0) of a radioactive source decreases to a value A after a time t according to the equation $A = A_0 e^{-\lambda t}$ where λ is the decay constant. Calculate A as a percentage of A_0 for a time of 100 s given that λ is 0.011 s^{-1}.

5 A parallel plate capacitor is made of two horizontal metal plates each having an area of 4.0×10^{-2} m^2 and separated by a distance of 1.5 mm. The potential difference between the plates is 500 V. Calculate:

(a) the capacitance of the capacitor
(b) the charge on a plate
(c) the energy stored in the capacitor.

(WJEC 2000, part question)

> HINT
>
> *Capacitance $C = \dfrac{\varepsilon A}{d}$ where ε is 8.86×10^{-12}, A is the area in m^2 of each plate and d is the plate separation in metres. 1.5 mm = 1.5×10^{-3} m. Charge q = CV and energy stored is $\frac{1}{2} CV^2$.*

6 A beam of electrons is directed at a target. They are accelerated from rest through 12 cm in a uniform electric field of strength 7.5×10^5 N C^{-1}.

(a) Calculate the potential difference through which the electrons are accelerated.
(b) Calculate the maximum kinetic energy in joules of one of these electrons.
(c) Calculate the maximum speed of one of these electrons.

(Edexcel 2001, part question)

(Electronic charge, $e = 1.60 \times 10^{-19}$ C and electronic mass, $m = 9.11 \times 10^{-31}$ kg)

> HINT
>
> *Electric intensity or field strength, $E = \dfrac{V}{d}$ where V is the potential difference across the distance d. Work done in accelerating an electron = gain of kinetic energy = e V where e is the electronic charge. Kinetic energy = $\frac{1}{2} m v^2$ where m is electron mass and v is electron speed or velocity.*

Answers to Test Yourself Questions

Exercise 4.1.1
1 (a) 10^2 (b) 10^6 (c) 10^4 (d) 10^5
2 (a) 1000 (b) 100 000 (c) 100 (d) 1000
3 (a) 9 (b) 8 (c) 16 (d) 32
4 (a) 10^6 or 64 (b) 2^5 (c) 10^8 (d) x^4
5 (a) 7000 (b) 74 000 (c) 332×10^3 (d) 213

Exercise 4.1.2
1 (a) 10^7 (b) 10^6 (c) 10^7 (d) 10^{12} (e) 2^{10}
2 (a) 2^2 or 4 (b) b^2 (c) p^2 (d) 7^2 or 49
(e) 4^2 or 16 (f) 4^2 or 16
3 (a) 2^6 or 64 (b) 8×10^{12} (c) a^{8b} (d) p^{2q}

Exercise 4.1.3
1 (a) 10^{-1} (b) 10^{-4} (c) 10^4
2 (a) 10^2 (b) 10^1 or just 10 (c) 10^3
3 (a) 10^4 (b) 10^{-2} (c) 10^{19}
4 (a) 10^4 (b) 10^{-7} (c) x^{-4}
5 (a) 100 (b) 9×10^{-6} (c) 7×10^{-2} (d) 2×10^{-2}

Exercise 4.1.4
1 (a) 4×10^7 or 400×10^5, etc. but preferably 40×10^6 (see chapter 7)
(b) 35×10^{-5} (c) 72×10^4 (d) 0.053×10^{-5}
(e) 8.0×10^{-5}

Exercise 4.1.5
(a) 10^8 (b) 2×10^2 (c) 25.65×10^{15} or 26×10^{15}
(d) 7.6×10^6

Exercise 4.1.6
(a) 12×10^{-5} (b) 7.5 (c) 6.5×10^3
(d) 6×10^4

Exercise 4.2.1
1 (a) 1024 (b) 729 (c) 486 (d) 39 (e) 12.8
(f) 5.657 or 5.7
2 (a) 0.19 (b) 0.0135 (c) 5×10^6 (d) 0.026
(e) 0.01 (f) 1.9×10^{-7} (g) 5.9

Exercise 4.2.2
1 0.051 newton
2 1053 or 1.0×10^3 watt

Exercise 4.3.1
1 (a) 20 000 or 2.0×10^4 disintegrations per second
(b) 10 000 or 10^4 or 1.0×10^4
(c) 2.8×10^4
2 0.165 or 0.17

Exercise 4.4.1
(a) 5.555×10^3 (b) 5.555×10^{-3} (c) 8.354×10^{-1}
(d) 9.54×10^8 (e) 4.35×10^{-2} (f) 2.82×10^{-3}
(g) 2.5×10^5

Chapter 5

Units

After completing this chapter you should:

- *know the rules for units*
- *know the symbols for the most frequently used quantities, multiples and submultiples*
- *have become used to SI units*
- *have learnt to convert other units to SI units*
- *understand the concept of dimensions*
- *be able to check formulae and units by considering their dimensions.*

5.1 The concept of units

How we use units

Measurements are usually made by counting and it is essential to know what is being counted, 400 metres for example. The metre is the *unit* for this measurement.
Also for the measurement to be useful its significance must be known, for example a race-track length = 400 metres.

Notice that, 400 metres, is really a product, namely a metre 400 times ... one metre, another metre, another ... counting up to 400.

KEY FACT   *A measurement is a product of a number and a unit.*

The *metre* is the *unit* of measurement (unit meaning one of many, an appropriate term because a typical measurement will be a number of them). The abbreviation for metre is m (see *Fig. 1*).

Other units commonly used in physics include the kilogram, second and ampere. In *Fig. 1* the N is the abbreviation for *newton*, the unit used for a *force* (i.e. for a push or a pull).

The 400 metre above was written with the unit in the singular, i.e. *metre* rather than *metres* not simply by personal choice but because it is accepted good practice. An s at the end of a hand-written unit could be poorly attached to the unit and mistaken for the symbol for second. However, it is unnatural and consequently often difficult to state units in the singular in conversation and when writing units in full in text. In calculations the rule should always be adhered to. This rule will be followed in all the remaining chapters of this book.

KEY FACT   *Track length = 400 metre or 400 m. Write units in the singular, metre not metres, cm not cms.*

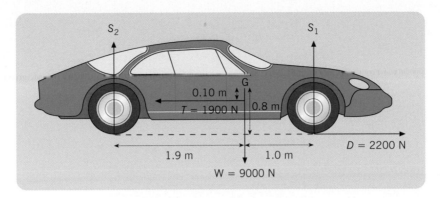

Fig. 1 *Units used in physics – a diagram typical of an A level physics examination paper.*

Now for the length of a two-lap race.

The race length = 2 × track length = 2 × 400 = 800 metre.

The 2 in the above equation is a *pure number* – it is simply a number not accompanied by any unit. The remaining part of the 2 × 400 m, the 400 m, is not a pure number but contains and needs the unit.

KEY FACT *A measurement must have its unit specified. Length = 2 × 400 m or 800 m. The 2 here is a pure number.*

Another important fact concerning units is that they can be *cancelled* just like numbers. This is noticed if you say that the distance covered in 5 hours at a speed of 3 km per hour is

$$\frac{3\ km}{1\ hour} \times 5\ hour$$ and equals 15 km. The hours have cancelled.

KEY FACT *Units can be cancelled like numbers.*

Remembering that a measurement consists of a number times its unit we have

$$\frac{measurement}{its\ unit} = \frac{number \times unit}{unit}$$ and cancelling gives

$$\frac{measurement}{its\ unit} = pure\ number$$

The term *quantity* can be used for measurement or number so $\frac{quantity}{unit}$ is a pure number.

In A level physics we like entries in tables of results to be pure numbers (as mentioned in Chapter 4) and the headings used should agree with this fact. For example $\frac{length}{metre}$ might be used or $\frac{acceleration}{m\ s^{-2}}$ (see *Fig. 3a* in Chapter 4).

Similarly in graphs the axes are marked with pure numbers and should be appropriately labelled, for instance with length/m or a/m s^{-2} (see *Fig. 2*).

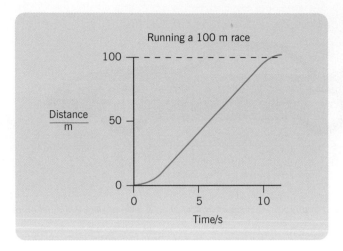

Fig. 2 *Labelling of a graph.*

The SI system

The **SI system** is (in French) the Système Internationale. This system is a set of rules and agreements, introduced many years ago, by a large number of nations so that their measurements and calculations could be more easily communicated between countries. The introduction of the system also allowed a great deal of simplification.

It assumes the use of decimal numbers, defines suitable units for scientific and other measurements and defines terms like *kilo* and *centi* that are used with units. The symbols to use as abbreviations for units are also specified.

Listed below are some examples of SI units and abbreviations for them.

Table 1 *SI System*

Quantity	SI unit	Symbol for unit
mass	kilogram	kg
length	metre	m
time	second	s
angle	radian	rad
electric current	ampere	A

A more complete list is given in Table 1 (page 225).

The SI unit for mass is not the gram, which you might have expected, but the kilogram. The SI unit for angle is not the familiar degree but the radian and this is explained in Chapter 12. (Note that the term *length* is often used in place of *distance*.)

The quantities mass, length and time are called *base quantities* and the SI units for these are **base units** because they are not defined in terms of any other quantities or units.

Historically, the kilogram was defined as the mass of a particular piece of metal kept in France and the metre as the length of another piece of metal also kept in France, but there are now other ways of defining these units.

(The other base units are the mole, kelvin, ampere and radian.)

KEY FACT *The metre, kilogram and second are all base units.*

Units other than base units are *derived* units. A simple example is the unit for measuring a speed or velocity, the *metre per second*. This unit like all other mechanics units can be derived from the mass, length and time base units. It is seen to be correct for velocity measurement because any velocity is a distance divided by a time. For example, 6 metres covered in 2 seconds means a velocity of 6 metres per 2 seconds or 3 metres per second.

$$\text{velocity} = \frac{\text{distance covered}}{\text{time taken}}$$

It will be a number of metres divided by seconds. The metre per second is best written as m s^{-1} but m/s is acceptable.

The SI unit for *acceleration* is m s^{-2}. We say 'metre per second squared' or m per s per s. The latter can be confusing and m per s^2 is better, m/s^2 is also acceptable. Writing m/s/s for metre per second per second should be avoided because it can be interpreted as $m \div \frac{s}{s}$ which equals m ÷ 1 and this equals just m.

The SI unit for force agrees with the formula *force = mass × acceleration* and is therefore kg × m × s^{-2} or kg m s^{-2} but it has been given a special name, the newton, abbreviation N.

KEY FACT *The m s^{-1}, the m s^{-2} and the newton are examples of derived units.*

Fig. 3 shows how some of the units used in mechanics are derived from base units. Some units are named after famous scientists such as Newton, Ampere, Volta and Ohm. When these names (sometimes shortened) are given to units they are spelt without a capital letter, for example, newton **not** Newton but the abbreviations are mostly capital letters, for example, N for Newton.

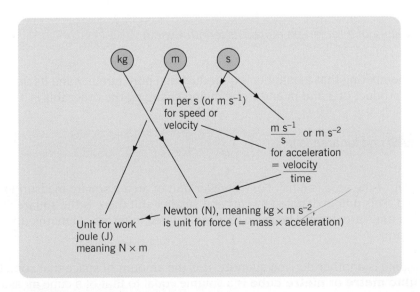

Fig. 3 Deriving mechanics units.

Multiples and *submultiples* of units make it easier to describe measurements. **kilo** denotes a thousand times and is placed in front of a unit to form a bigger unit such as the kilometre. So this prefix forms a *multiple* of the original unit.

The abbreviation for kilo is k so that km means kilometre and 1 kilometre (km) = 1000 m.

Other prefixes for units include mega (symbol M) meaning *a million times* and are listed in Table 2 (*page* 227).

Submultiples of units have prefixes that form smaller units. Examples of these are **centi** meaning hundredth of (for example, cm for centimetre), **milli** for thousandth of (e.g. mm for millimetre) and **micro** for millionth of (abbreviation μ as in μA for microampere).

$$1 \text{ km} = 1000 \text{ m} \qquad 1 \text{ cm} = 10^{-2} \text{ m} \qquad 1 \text{ mm} = 10^{-3} \text{ m} \qquad 1 \text{ } \mu\text{m} = 10^{-6} \text{ m}$$

No gap should be left between a prefix and the unit it applies to. So millisecond has the abbreviation ms and is not to be confused with m s (with a gap) which would indicate metre second. The abbreviation g ms^{-1} means gram per millisecond while g m s^{-1} means gram metre per second.

Some quantities use units that have not been given special names. Area and volume are examples. *Area* means amount of surface, and the area of a square surface measuring 1 metre by 1 metre is called a **metre square** or a **square metre**. This is the SI unit for area and its abbreviation is **m^2**. Any other area can be measured by counting the number of metre squares that would be needed to cover that area. For a rectangular area having a length L and a width w the area equals $L \times w$ (*see Fig. 4*).

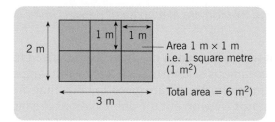

Fig. 4 *Area of a rectangle.*

KEY FACT   *The area of a rectangle equals its length × width.*

Remembering that a quantity is a product of a pure number and its unit an area like 3 m × 2 m is a product of 3 and m and 2 and m and changing the order this is $3 \times 2 \times m \times m$ or 6 m^2.

KEY FACT   *The SI unit for area is m^2.*

In *Fig. 4* the area is 6 m^2. For a smaller unit of area a square measuring 1 cm by 1 cm can be used. This unit is denoted by cm^2 and smaller still there is the square millimetre, mm^2. These area units are derived because they are defined in terms of lengths (the length and width of a square).

Volume is another derived quantity. It is the amount of space that something occupies. A **cubic metre** or **metre cube** is a volume equal to that of a cube measuring 1 m by 1 m by 1 m. This is the SI unit for volume and is denoted by **m^3**.

Any volume can be measured by counting the number of metre cubes that would be needed to fill the volume. For a rectangular volume (i.e. a *cuboid*) shown in *Fig. 5* the volume is 18 m³

The volume of a rectangular block (a cuboid) is given by the formula

$V = Lwd$ where L = length, w = width and d = depth.

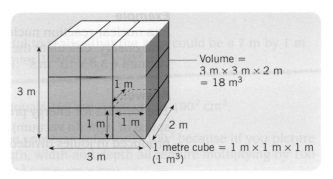

Volume =
3 m × 3 m × 2 m
= 18 m³

1 metre cube = 1 m × 1 m × 1 m
(1 m³)

Fig. 5 *Volume of a cuboid.*

KEY FACT The SI unit for volume is m³.

Conversion of units

A simple example of converting units is changing 100 cm to become 1.00 m. The cm measurement has to be divided by 100 or multiplied by $\frac{1}{100}$. These multiplying figures are called *conversion factors.*

Formulae met in A level physics should all work correctly if they are used with SI units. So if a quantity is of known size and you want to use it in a calculation you should always ensure that its units are SI units unless you are certain that the conversion factors you would use would simply cancel out during the calculations.

KEY FACT Be safe use SI units in calculations.

Suppose that a distance is 0.2 km and this length is to be converted to a length in metres.

Remembering that km means kilometre which is *1000 times* a metre, the 0.2 km becomes 0.2×1000 m. Similarly 7.3 mm = $7.3 \times \frac{1}{1000}$ m = 0.0073 m.

Conversions like these apply also to measurements other than lengths.

HINT   *You might find it difficult to decide whether a conversion requires a multiplication or a division. I want to convert 1200 mm to metres. I know that mm means a thousanth of a metre but do I multiply the 1200 by a thousand or do I divide by a thousand to get the length in metres? This is a common problem and the best approach is to ask oneself whether the conversion is into a greater number of smaller units or into larger units so that there are fewer of them. So converting the 1200 little millimetres into much bigger metres will give a smaller number than 1200 and the 1200 must be divided by 1000 to give 1.200 m.*

KEY FACT Converting to larger units gives a smaller number, for example 8 mm = $\frac{8}{1000}$ m = 8×10^{-3} m or 0.008 m.

Chapter 6

Roots of numbers

After completing this chapter you should:

- *understand the concept of roots and the use of the $\sqrt{\ }$ sign*
- *be able to obtain values of roots from your calculator*
- *know how to express roots as powers of ten*
- *be able to use equations containing roots.*

6.1 Roots

The meaning of this term

You might be calculating the frequency of a vibration or an alternating electric current and want the value of $\sqrt{4.0 \times 10^4}$.

The answer is 2.0×10^2 as will soon be explained.

$2 \times 2 \times 2 \times 2$ equals 16 or 2^4 and the process that gives the 2^4 is called 'finding the fourth power' of 2. The reverse of this process starts with 2^4 or 16 and produces an answer of 2. This process is called finding (or taking) the *fourth root* of the 16. A **root** of a number is a value which can multiply by itself to produce the number. So 3 is a root of 243 because 3s can multiply to give 243. In fact $3 \times 3 \times 3 \times 3 \times 3 = 243$ and because five threes are needed 3 is the fifth root of 243.

Now it should be noticed that $-2 \times -2 \times -2 \times -2 = 16$. This means that -2 is also an answer for the fourth root of 16.

> The fourth root of 16 is 2 or −2.

The symbol for root is $\sqrt{\ }$ and the symbol for 'the fourth root of' is $\sqrt[4]{\ }$. The 4 should be written close to the $\sqrt{\ }$ sign, ideally above the foot of the sign.

$$\sqrt[4]{16} = 2 \text{ or } -2$$

Similarly

$$\sqrt[2]{9} = \sqrt[2]{3 \times 3} \qquad \text{or} \qquad \sqrt[2]{-3 \times -3} = 3 \text{ or } -3$$

$$\therefore \ \sqrt[2]{9} = 3 \text{ or } -3$$

The $\sqrt{\ }$ sign must be written so that the whole of the expression it refers to is covered by the sign. So $\sqrt[4]{16 \times x^4} = 2x$ or $\sqrt[4]{(16 \times x^4)} = 2x$ is correct but $\sqrt[4]{16} \times x^4 = 2x$ is *not* correct. (In fact $\sqrt[4]{16} \times x^4 = 2 \times x^4$ (or just $2x^4$) because the root sign applies only to the 16.)

The second root is always called the **square root** and is denoted by just $\sqrt{\ }$ instead of $\sqrt[2]{\ }$.

KEY FACT $\sqrt{\ }$ *means* $\sqrt[2]{\ }$

One example of a square root being involved in a physics calculation is the formula

$T = 2\pi \sqrt{\dfrac{L}{g}}$ for the period of a simple pendulum (*see Fig. 1*). In this formula there is a small

gap between the 2π and the $\sqrt{}$ sign. The 2π is simply multiplying the $\sqrt{\dfrac{L}{g}}$, not indicating that the root is something other than a square root.

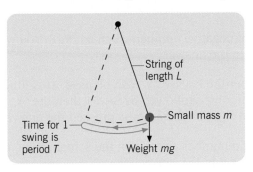

Fig. 1 *A simple pendulum.*

For $^3\sqrt{27}$ we are looking for three identical factors that multiply together to make the given number 27. They are of course threes.

$$^3\sqrt{27} = 3$$

Because an odd root is required ($^3\sqrt{}$) of a positive number (27) there is only one root. This is +3. You cannot have −3 as an alternative answer because $-3 \times -3 \times -3 = -27$ not 27.
For 'third root' the term **cube root** is always used. There are no special names for fourth root upwards.

   The cube root of 27 is 3.

Example
Calculate the current (I) flowing through an inductor having an inductance (L) of 0.50 henry when the energy (E) stored in the inductor is 1.00 joule.

Answer
The formula for the energy stored is $E = \frac{1}{2}LI^2$.

   $\therefore$ $1.00 = \frac{1}{2} \times 0.5 \times I^2$ and this means that $1.00 = \frac{1}{4}$ of I^2 and 4.00 must equal I^2.

Consequently $I = \sqrt{4}$ i.e. $I = 2$. The unit for I is ampere. $I = 2.0$ A.
(Why 2.0 A not just 2 A? This will be explained in Chapter 7.)
$I = -2.0$ A is an alternative answer (current flowing in the opposite direction).

Roots on a calculator

On your calculator there is a key marked with $\sqrt{}$ for finding the square root of a number. When a number is already displayed in the answer part of the calculator's screen, then pressing the $\sqrt{}$ key followed by = gives the square root of that answer. If the screen has been cleared then the $\sqrt{}$ key can be pressed before entering the number whose root is required.

So

 gives 3

or

$\boxed{AC}\boxed{\sqrt{}}\boxed{9}\boxed{=}$ gives 3

Unfortunately a calculator will give only one root, for instance 2 for $\sqrt{4}$ not −2.

Your calculator also has a key $^3\sqrt{}$ key this can be used for a cube root in the same way that the square root key was used.

It may not like to find roots of negative numbers such as $^3\sqrt{-8}$. You would need to find $^3\sqrt{8}$ and insert the − sign needed in the answer.

For the fourth root of a number the key for $^x\sqrt{}$ can be used. This function requires the SHIFT key to be pressed first. Try entering 4 followed by SHIFT then the $^x\sqrt{}$ key and then enter 16 because you want $^4\sqrt{}$ of the 16. When the = key is pressed the answer given should be 2.

If you want the fourth root of a number already on the screen as an answer then the x^y key could be used for $^4\sqrt{16}$ because it is $16^{\frac{1}{4}}$ (see below).

Example

A small mass m is moving at constant speed in a horizontal circular path of radius R due to a force F acting on the mass and directed towards the circle's centre. If $F = 200$ newton, $R = 1.00$ metre, $m = 1.00$ kg, calculate v from the equation $F = mv^2/R$. (In Chapter 8 we see that this equation can be rearranged to become $v^2 = \frac{FR}{m}$.)

Answer

$$v^2 = \frac{F \times R}{m} = \frac{200 \times 1}{1} = 200$$
$$\therefore\ v = \sqrt{200}$$

and by use of a calculator $v = 14.14$.

The unit for v is metre per second and $v = 1.4 \times 10^1$ m s^{-1} or 14 m s^{-1}.

Test Yourself

Exercise 6.1.1

Use a calculator to obtain the positive square root of

1 (a) 25   (b) 36   (c) 144   (d) 16.33   (e) 2.2   (f) 2
    (g) 100   (h) 400   (i) 200   (j) 10   (k) 40

HINTS AND TIPS   *Shorten any long answers to four figures but see Chapter 7.*

Example

Calculate the length L of each side of a cube that is made of a substance having a density of 250 kg per cubic metre and weighs 31.25 kg.

Answer

The equation to use here is $D = m/V$ (density = mass/volume) or $V = m/D$. For the cube the volume $= L^3$ so we have

$$L^3 = \frac{m}{D} = \frac{31.25}{250} = 0.125$$

The cube root of 0.125 is found from a calculator to be 0.50 and the answer is $L = 0.50$ metre.

Shorten long answers to four figures.

1 Use the $^x\sqrt{\ }$ and $^3\sqrt{\ }$ keys of your calculator to determine the cube root of each of the following numbers. Use one key to get the answer, then check using the other key:

(a) 125   (b) 10 000   (c) 1000   (d) 27.1   (e) 1   (f) 1.12   (g) 9.1
(h) 8000

2 Find the cube root of each of these numbers, use a calculator or another method:

(a) 64   (b) −27   (c) −27.5   (d) −1000   (e) −1/1000

6.2 Squaring to remove square roots

In an equation the quantity to be evaluated may be within a square root sign. Squaring both sides of the equation may be appropriate. For example $\sqrt{x} = 3$ becomes $x = 3^2 = 9$.

KEY FACT   *Squaring both sides of an equation may remove a square root.*

Example

A mass m hangs on the lower end of a vertical spring and is then disturbed and allowed to move up and down before it eventually becomes still. The period T of this oscillation is 2.00 seconds and the mass m is 0.150 kg. Given that $T = 2\pi\sqrt{m/k}$ where k is the *spring constant*, calculate m/k.

Answer

$T = 2\pi\sqrt{\dfrac{m}{k}}$. Squaring both sides of this equation gives

$$T^2 = 4\pi^2 \times \frac{m}{k} \text{ so that}$$

$$2^2 = 4 \times 9.872 \times \frac{m}{k}$$

or $4 = 40 \times \dfrac{m}{k}$ approximately,

giving $I = 10 \times \dfrac{m}{k}$ and $\dfrac{m}{k} = \dfrac{1}{10}$ or 0.100.

The unit for m/k is kilogram metre per newton (kg m N^{-1}) (See Chapter 5, exercise 5.2.1, question 3.)

Find the value of the unknown quantity:

1 (a) $\sqrt{x} = 5$ (b) $\sqrt{x+3} = 3$ (c) $\sqrt{5x} = 7$ (d) $\sqrt{\dfrac{4}{x}} = 2$

6.3 Powers for roots

One of the aims of this chapter is to explain how $\sqrt{4 \times 10^4} = 2 \times 10^2$. The *power* 10^4 has changed to 10^2 because of taking the square root.

In Chapter 4, the relationship $x^a \times x^b = x^{a+b}$ has been met.

So $x^{\frac{1}{2}} \times x^{\frac{1}{2}} = x^1 = x$ but $\sqrt{x} \times \sqrt{x} = x$ and it is clear that $x^{\frac{1}{2}}$ or $x^{0.5}$ is the same as $\sqrt{x}$ or $\sqrt{x^1}$.

KEY FACT $\sqrt{x} = x^{\frac{1}{2}}$
Similarly $x^{\frac{1}{3}} \times x^{\frac{1}{3}} \times x^{\frac{1}{3}} = x^1$ showing that $x^{\frac{1}{3}}$ is $\sqrt[3]{x}$.

Clearly, in general,

KEY FACT $x^{\frac{1}{n}} = \sqrt[n]{x}$.

It has also just been stated that $\sqrt{x^1} = x^{\frac{1}{2}}$ and this illustrates that to obtain the square root of a power you halve the index. Similarly for a cube root you divide the index by 3.

KEY FACT *To find the nth root of a power divide the index by n, for instance $\sqrt[3]{10^6} = 10^2$.*

If you enter $2 \times 8 =$ on your calculator to display 16, you can then press the x^y key followed by 0.25 and $=$ to get an answer of 2 as expected. You have obtained $16^{\frac{1}{4}}$ which is $16^{0.25}$.

Test Yourself

Exercise 6.3.1

Express each of the following roots in the form x^p, e.g. for $\sqrt{x}$ the answer is $x^{\frac{1}{2}}$ or $x^{0.5}$.

1 (a) $\sqrt[5]{x}$ (b) $\sqrt[4]{a}$ (c) $\sqrt[3]{x^2}$ (d) $\sqrt[3]{x^3}$ (e) $\sqrt[2]{x^4}$ (f) $\sqrt[4]{x^2}$

Example
The volume per minute Q of liquid that flows out of a tube with a uniform bore depends upon the radius r of the bore and $Q = Cr^4$ where C is a constant. If initially $Q = 5$ millilitres per minute but is subsequently reduced to 2 millilitres per minute by using a tube with radius $\dfrac{r}{F}$, what is the value of F?

Answer

At first $Q = Q_1 = 5 = Cr^4$. Subsequently $Q = Q_2 = 2 = C \times \left(\dfrac{r}{F}\right)^4 = C \times \dfrac{r^4}{F^4}$

Dividing the equations we get

$$\frac{Q_1}{Q_2} = \frac{C \times r^4}{C \times r^4 / F^4} = \frac{C \times r^4 \times F^4}{C \times r^4} \text{ (by multiplying top and bottom by } F^4\text{)}$$

and cancelling gives $\dfrac{Q_1}{Q_2} = F^4$ and so $F^4 = 5/2$ or 2.5 and $F = 2.5^{\frac{1}{4}}$ or $2.5^{0.25}$.

Using a calculator and as an example its x^y key, $x = 2.5$ and $y = 0.25$ and the answer is 1.26.

Test Yourself

Exercise 6.3.2

For each of the following equations find the positive value of x that fits, i.e. solve each equation for x.

1 (a) $x^3 = 8$ (b) $x^4 = 16$ (c) $x^3 = 12\ 167$
  (d) $2x^3 = 16$ (e) $5x^6 = 15$ (f) $(2x)^3 = 512$

6.4 Roots $\sqrt{a \times b}$ and $\sqrt{\dfrac{a}{b}}$

If you were calculating the speed of an aircraft you might want to work out an expression like $\sqrt{4.0 \times 10^4}$ and could use the rule $\sqrt{a \times b} = \sqrt{a} \times \sqrt{b}$ to get 2.0×10^2. The rule is easily proved by writing

$$\sqrt{a \times b} \text{ as } \sqrt{\sqrt{a} \times \sqrt{a} \times \sqrt{b} \times \sqrt{b}} \text{ and this equals } \sqrt{a} \times \sqrt{b}$$

KEY FACT $\sqrt{a \times b} = \sqrt{a} \times \sqrt{b}$.

You could easily prove the rule for cube roots or other roots.

For the speed of the aircraft:

$$\sqrt{4.00 \times 10^4} = \sqrt{4.00} \times \sqrt{10^4} = 2.00 \times 10^2 = 200$$

Using the rule for such a calculation is a little quicker than using the EXP key on your calculator for the 10^4.

Simplifying an expression like $\sqrt{4.00 \times 10^7}$ illustrates a neat trick The power of ten can be changed to an even power ready for the square root of this power to be found by halving the exponent. The other part of the number, the 4.00 is adjusted to correct for the power change made, i.e. the 4.00 is changed to 40.0 if the 10^7 is changed to 10^6 and a calculator is only needed for finding the root of this corrected number. So

$$\sqrt{4.00 \times 10^7} = \sqrt{40.0 \times 10^6} = \sqrt{40.0} \times \sqrt{10^6} = 6.3 \times 10^3$$

This procedure is useful when you are making a rough check. You would expect $\sqrt{40}$ to be about 6 and the answer to be about 6×10^3.

Be careful! The rule for the root of a product does *not* apply to a sum or difference. For example $\sqrt{a^2 + b^2}$ is NOT equal to $a + b$ (because $(a + b)^2 = a^2 + b^2 + 2ab$ as explained in Chapter 3).

An expression of the form $\sqrt{\dfrac{a}{b}}$ is seen for example in the formula for the period of vibration of a mass on a spring $T = 2\pi \sqrt{\dfrac{m}{k}}$ and the rule $\sqrt{\dfrac{a}{b}} = \dfrac{\sqrt{a}}{\sqrt{b}}$ could be useful when working with a formula like this.

You might like to test the rule yourself using simple numbers.

Example

If the volume of a unit cell in a sodium chloride lattice is given as 2.24×10^{-29} cubic metres, what is the length of each side of this cube?

Answer

If L represents the length of a side, then the volume of the cube $= L^3$. So for the unit cell $L^3 = 2.24 \times 10^{-29}$.

On your calculator you use

$$\boxed{2}\,\boxed{.}\,\boxed{2}\,\boxed{4}\,\boxed{\text{EXP}}\,\boxed{-}\,\boxed{2}\,\boxed{9}\,\boxed{=}$$

to enter the value of L^3 and then press the $\sqrt[3]{}$ key get an answer of 2.8×10^{-10} metre.

Test Yourself

Exercise 6.4.1

Obtain the positive values of the following roots.

1 (a) $\sqrt{4 \times 10^8}$ (b) $\sqrt{8 \times 10^2 \times 2 \times 10^2}$
   (c) $\sqrt{90 \times 10^3}$

HINT *Consider $\sqrt{9 \times 10^4}$.*

   (d) $\sqrt[3]{27 \times 10^9}$ (e) $\sqrt[3]{2.7 \times 10^{10}}$ (f) $\sqrt{4 \times 10^{-8}}$ (g) $\sqrt{0.4 \times 10^{-7}}$

   (h) $\sqrt{0.0016 \times 10^{-4}}$ (i) $\sqrt{\dfrac{80 \times 10^{-2}}{5 \times 10^6}}$

Test Yourself

Exercise 6.4.2

Simplify:

1 (a) $\sqrt{\dfrac{9}{x^2}}$ (b) $\sqrt[3]{\dfrac{27}{8b^3}}$

Give only positive answers here, negative answers are considered below.

   (c) $\sqrt{4a^2b^2}$

HINT *Remember $\sqrt{ab} = \sqrt{a} \times \sqrt{b}$ and $\sqrt{\dfrac{a}{b}} = \dfrac{\sqrt{a}}{\sqrt{b}}$.*

6.5 Positive and negative roots

It has been mentioned already in this chapter that multiplying two negative numbers gives a positive product and consequently a positive number, like 4, has two square roots, namely +2 and −2. Also the cube root of −27 is −3.

You should recall that

● an even root of a + number has two values,

● an even root of a – number does not exist,

● an odd root of a negative number has one negative value.

In A level physics calculations it is the two values for the square root of a positive number that are most important, but you may want to try exercise 6.5.1 in which other roots are featured. Usually only one of the square roots is important, the other one giving an answer such as a negative absolute temperature that makes no sense for the question.

Graphs showing roots and powers of numbers are shown in *Fig. 2*.

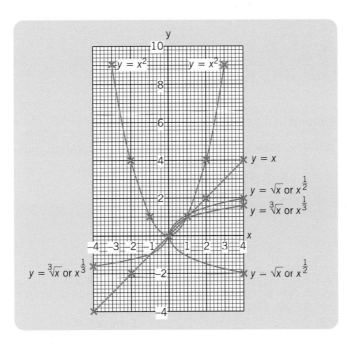

Fig. 2

Test Yourself

Exercise 6.5.1

1 How many roots (1 or 2 or none) are there for each of the following:

(a) $^2\sqrt{4}$ (b) $\sqrt{-4}$ (c) $\sqrt{9}$ (d) $^3\sqrt{-3}$

(e) $^3\sqrt{4}$ (f) $^5\sqrt{32}$ (g) $^5\sqrt{-6.7}$ (h) $\sqrt{\dfrac{1}{4}}$

2 Write down all the + and – roots for each of the following, writing 'none' if there are none. Use a calculator as necessary.

(a) $^5\sqrt{32}$ (b) $^3\sqrt{-125}$ (c) $\sqrt{-6.4}$ (d) $\sqrt{1}$

(e) $\sqrt{5.7}$ (f) $^3\sqrt{-3.9}$ (g) $^7\sqrt{777}$ (h) $\sqrt{10}$

Exam Questions

Exam type questions to test understanding of Chapter 6

Exercise 6.5.2

1 The period (T seconds) of oscillation of a mass at the end of a spring is given by the formula $T = 2\pi \sqrt{\dfrac{e}{g}}$. The e is the extension of the spring when the object is at rest. Calculate T for $e = 0.08$m metre given that $g = 9.81$ m s^{-2}.

2 Stefan's law relates the power (P) radiated by a hot surface to the absolute temperature (T) of the surface. $P = \sigma A T^4$. A is the area of the surface and σ is the Stefan constant. Calculate T for $P = 100$ watt given that σ for the surface is 5.7×10^{-8} W m^{-2} K^{-4}, and that the surface has an area of 25×10^{-4} m^2. (All units SI. The equation can be rewritten as $T = \sqrt[4]{\dfrac{P}{\sigma A}}$. The unit for T is the kelvin (K).)

HINT $\quad T^4 = \dfrac{P}{\sigma A}.$

3 The speed of sound in air at 0 °C is given by the formula $c = \sqrt{\dfrac{\gamma P}{\rho}}$ where γ is 1.4, ρ is 1.29 kg m^{-3} and P is 1.0×10^5 Pa. Calculate this speed. (All SI units.)

HINT $\quad$ *Use EXP key or write* $\sqrt{\dfrac{14}{1.29}} \times 10^2.$

Answers to Test Yourself Questions

Exercise 6.1.1
1 (a) 5 (b) 6 (c) 12 (d) 4.041 (e) 1.483 (f) 1.414 (g) 10 (h) 20 (i) 14.14 (j) 3.16 (k) 6.3

Exercise 6.1.2
1 (a) 5 (b) 21.54 (c) 10 (d) 3.003 (e) 1 (f) 1.038 (g) 2.088 (h) 20
2 (a) 4 (b) −3 (c) −3.018 (d) −10 (e) −1/10

Exercise 6.2.1
1 (a) 25 (b) 6 (c) 9.8 (d) 1

Exercise 6.3.1
1 (a) $x^{\frac{1}{5}}$ or $x^{0.2}$ (b) $a^{1/4}$ or $a^{0.25}$ (c) $x^{2/3}$ (d) x^1 or x (e) x^2 (f) $x^{\frac{1}{2}}$

Exercise 6.3.2
1 (a) 2 (b) 2 (c) 23 (d) 2 (e) 1.201 (f) 4

Exercise 6.4.1
1 (a) 2×10^4 (b) 4×10^2 (c) 3×10^2 (d) 3×10^3 (e) 3×10^3 (f) 2×10^{-4} (g) 2×10^{-4} (h) 4×10^{-4} (i) 4×10^{-4}

Exercise 6.4.2
1 (a) $\frac{3}{x}$ (b) $\frac{3}{2b}$ (c) $2ab$

Exercise 6.5.1
1 (a) 2 (b) 0 (c) 2 (d) 1 (e) 1 (f) 1 (g) 1 (h) 2
2 (a) 2 (b) −5 (c) None (d) +1, −1 (e) ±2.387 (f) −1.574 (g) 2.588 (h) ±3.162

Chapter 7

Errors

After completing this chapter you should:

- *know how the possible error in a measurement can be recorded*
- *understand the importance of significant figures*
- *be aware of errors combining when a calculation involves a number of measurements*
- *know simple ways of allowing for possible errors when a result is calculated.*

7.1 Experimental errors

When a length is measured the difference between the measured length and the true length is the **error**. Accurate measurements are those with small errors. The causes of errors may be limitations of measuring equipment (*instrument errors*) or of the experimenter's skill. Whenever a measurement is made it should be assumed that at least some small error is included in the recorded value.

Imagine a measurement made with a metre rule held in a very shaky hand. The measurement may be greater or less than the true value and may be small or large. It is a **random error**. Repeating such measurements gives a spread of answers because the **precision** is low, but if we assume the correct answer is somewhere in the middle of the group of answers we can get close to the correct value as explained in Chapter 12.

The limitations of measuring equipment can cause errors. Taking a metre rule as our example again, the graduations on the rule are no closer than one millimetre apart and so readings can only be made to the nearest millimetre mark. The length recorded may therefore be greater or less than the true length. It would be unreliable to estimate a reading between adjacent millimetre marks. The nearest mark is used.

If a value of 27.4 cm is recorded this will indicate a length that is closer to 27.4 than to 27.3 or to 27.5. A length half way between 27.3 and 27.4 should be recorded as 27.4 cm so that a half-way length is always taken as the next highest reading.

27.4 means 27.35 up to but not including 27.45

Measurements other than lengths, for example, currents read from meters, and temperatures from thermometers, which are read from a scale, will also include small errors. These errors are avoided if a digital meter is used, but accuracy is limited by the number of digits displayed.

Due to the same causes of possible error a repeat measurement may be a slightly different value, a metre rule perhaps being read as 27.3 and next time as 27.4 cm.

The above discussion showed that metre rule measurements involving lengths could be smaller than the recorded value by up to half a millimetre or greater by such an amount. Unless a mistake is made by the person reading the scale the length cannot have errors greater than this + or − 0.5 mm (i.e. ±0.5 mm is the **maximum possible error** or simply the *possible error*).

The result of a calculation may contain errors due to causes other than those of measurements used in the calculation. The calculation may use a value for π (*pi*) such as 3.142 and this is not quite correct. 3.141592 7 is better but still not perfect. Even a calculator will introduce errors and will for example divide 2 by 6 and give 3.333×10^{-1} or 0.3333 or 0.3333333 whereas 0.3333333333 would be more correct.

Very likely a length measured on a metre rule will have an error less than the maximum possible error. So the *probable* error will be less than the *possible* error. In the case of radioactivity measurements, for example, the experimental errors are largely due to the random nature of the radiations being counted and a probable error or an RMS error (*see Chapter 12*) is considered but generally, values given for errors should be assumed to be the maximum possible values.

Recording possible errors

One example of recording possible error was the example in which the 27.4 cm measurement meant that the true length could be from 27.35 up to almost 27.45 cm. Another way of displaying the possible error in this measurement emphasises the error by writing it with the recorded value, for example, as 7.4 ± 0.05 cm.

   27.4 cm means 27.4 ± 0.05 cm

It is unfortunate here that 27.4 ± 0.05 includes 27.45 and we really would like it to mean *almost* 27.45 (because we have said that the higher figure of 27.5 would include 27.45 as one of its possible values). You may need to remember the *almost*.

A third way of describing a possible error is to show it as a percentage and this is explained after exercise 7.1.1.

The *symbol* > means **is greater than** and < means **is less than** so for a length (*L*) of 3.5 ± 0.05 mm we can write 3.55 > *L* > 3.45.

HINT
> ≫ *means is much greater than and* ≪ *indicates is much less than.*

Example
What is the greatest and smallest power measured as 23 watt?

Answer
Half-way between 23 and 24 is 23.5 = 23 + 0.5. The greatest power implied by 23 W is almost 23.5. The smallest is 23 − 0.5 = 22.5 W.

Test Yourself Exercise 7.1.1

1 What are the greatest and smallest possible lengths for each of the following readings?

(a) 34.3 cm   (b) 34.37 cm   (c) 9.9 mm   (d) 2.0 m   (e) 50.99 mm

2 What reading ideally should be obtained on an ammeter with a scale having only ampere and tenth of ampere markings if a more accurate meter reads the current as:

(a) 1.74 A   (b) 2.17 A   (c) 1.45 A

HINT   *Choose nearest marking.*

(d) 1.30 A   (e) 2.00 A   (f) 2.01 A   (g) 4.05 A   (h) 9.95 A

Percentage errors

Percentages were explained in Chapter 2. Possible errors are often stated as percentages of the recorded value. Since 0.05 cm is 1.825% of 27.4 cm ($100 \times \dfrac{0.05}{27.4} = 1.825$) the possible error can be shown by writing $27.4 \pm 2\%$. For A level physics you need to state percentage errors only to the nearest whole number unless asked to do otherwise.

An error described as ± 0.5 mm is an **absolute error**. A percentage is not *absolute* but **relative**, because the error is compared with something, namely the measured value.

In many experiments of the kind carried out in A level science courses a good measurement will have no more than $\pm 1\%$ possible error.

Example
Express the error in the measurement of 4.52 ± 0.05 cm^3 as a percentage error.

Answer
0.05 as a percentage of 4.52 is $100 \times \dfrac{0.05}{4.52}\%$ or 1.106% and the answer required is $\pm 1\%$.

Test Yourself Exercise 7.1.2

1 Rewrite each of the following measurements with the possible error shown as a percentage instead of an absolute error.

(a) 2.5 ± 0.05 volt   (b) 300 ± 30 (c) 6.25 ± 0.005 cm^3

2 Rewrite each of the following measurements with the possible error displayed as an absolute value instead of a percentage.

(a) $35 \pm 10\%$ (b) $72 \pm 1\%$ (c) $4.0 \pm 1.25\%$ (d) $9.0 \pm 3\%$

3 Which of the following three recorded measurements has the smallest percentage error?

(a) 9.99 ± 0.005 cm   (b) 4.44 ± 0.005 cm   (c) 1.11 ± 0.005 cm

HINT   *Could calculate each percentage error but same absolute error (0.0005) gives highest percentage in smallest measurement.*

7.2 Significant figures

A simple and quick way to specify possible error has been to use an appropriate number of figures in the measurement recorded. 27.4 cm indicates that a useful figure beyond the 4 cannot be recorded. So 27.43 cm will never be obtained from an ordinary metre rule. A length very close to 27 will be recorded as 27.0, being nearer to this length than to 27.1 or to 26.9. The zero here serves a purpose, it is *significant*, 27.0 telling us more than 27. The number 27.4 consists of **three significant figures** and has three significant figure accuracy. The measurement is recorded to **3 s.f**.

When *scientific mode* is selected on your calculator the number of significant figures required in the display has to be entered.

Note that in a number like 0.054 the zero following the decimal point is not significant in the way explained above because it can be regarded as acting simply as a spacer between the 5 and the decimal point. It shows where the significant part of the number starts, namely with the 5 in the hundredths position.

The zero before a decimal point, for example, in 0.67, serves no purpose other than to emphasise the presence of the decimal point.

KEY FACT *In the number 0.060 only the 6 and the right-hand zero are significant. The number has 2 significant figures.*

Test Yourself

Exercise 7.2.1

1 How many significant figures are there in each of the following numbers?

   (a) 3.47   (b) 2.30   (c) 0.3774   (d) 1.056   (e) 256

2 Write down the absolute error implied by each of the following measurements.

   (a) 2.1 cm   (b) 2.15 m   (c) 2.162 m^2

A difficulty overcome

If a measurement indicates a value within 10 of say 200 it could be misleading just to quote 200. This number appears to have 3 significant figures. The right-hand zero of this would suggest that the value is closer to 200 than to 199 or 201 rather than somewhere between 190 and 210. Really this zero is acting as a spacer to ensure that the 2 tells us how many hundreds. This difficulty is overcome by showing the possible error as 200 ± 10 or $200 \pm 5\%$.

Another way of dealing with such a number is to write the number using **standard form** as explained in Chapter 4. The 200 is written as 2.00×10^2. Now the 2.0×10^2 (in which the 2.0 has 2 s.f.) implies a range of possible value from 1.95×10^2 to almost 2.05×10^2 or 195 to 205. We choose the 2.0×10^2 because it gives a reasonable indication of the possible error although not the 190 to almost 210 that would be preferable. The 2.00×10^2 would imply a value between 199.5 and almost 200.5 and therefore too great an accuracy.

KEY FACT 2.0×10^2 *is almost the same as* 200 ± 10.

A large number written for example as 654 321 is not likely to be as accurate as it looks and standard form allows it to be stated with appropriate accuracy.

Example
What is the possible error implied for a force of 8.36×10^5 newton?

Answer

8.36×10^5 N $= (8.36 \pm 0.005) \times 10^5$ N.

The possible error is therefore $\pm 0.005 \times 10^5$ or ± 500 N.

Test Yourself

Exercise 7.2.2

What possible error is implied in each of the following standard form numbers?

1 (a) 2.54×10^3 (b) 3.5×10^4 (c) 3.444×10^3 (d) 2.4×10^6

7.3 Combining errors

Possible errors imply uncertainty about the true values of the quantities to which they relate and in view of this uncertainty there is no point in getting very accurate values for calculated errors.

In many calculations the quantities concerned multiply together or divide but do not add or subtract. Calculating the heat supplied by an electric heater may use the formula $Q = P \times t$ (heat = power $\times$ time). Calculating a resistance may use $R = V/I$.

KEY FACT   *Add the percentage errors of the quantities involved to get the percentage error in the product.*

This is a **rule** and is explained in Appendix I.

Suppose that a rectangular area has sides of 5.00 cm and 7.00 cm but was measured as 5.04 cm and 7.06 cm. The area would be calculated as $5.04 \times 7.06 = 35.58$ cm^2 instead of 35.0 cm^2, an absolute error of $0.58 = 0.6$ cm^2 or a percentage error of $\frac{0.58}{35} \times 100 = 1.66\%$.

The percentage error for the 5.04 is $\frac{0.04}{5.04} \times 100$ which is 0.8% and for the 7.06 is $\frac{0.06}{7.06} \times 100$ or 0.85%.

Adding these percentage errors gives 1.65%. This result is close to the expected 1.66% and so closely agrees with the adding of percentages rule.

If for this area the measurements were described as 5.00 ± 0.04 cm and 7.00 ± 0.06 cm then the lowest possible value for the area would be 34.42 cm^2 and the percentage error works out to be −1.6%. Note that 5.00 + 0.04 multiplied by 7.00 − 0.06 would give a smaller error. We want the maximum possible error and this is +1.6% or −0.6%, i.e. ±1.6%. This rule is used in the next example.

HINT ▷ *Adding the absolute errors of +0.04 and +0.06 would give an absolute error of +0.1. This would not agree with the +0.58 cm² error and as a percentage of the 35.58 would give +0.003% which is also wrong.*

Adding % errors also applies to divsions. For example $\dfrac{7 \pm 0.01}{7 \pm 0.01}$ could be as high as $\dfrac{7.01}{6.99}$ which is 1.003 and as low as $\dfrac{6.99}{7.01}$ which is 0.997, i.e. $7 \pm 0.3\%$. This agrees closely with the 0.14% error of the 7 ± 0.01 numerator being added to the 0.14% error of the denominator.

KEY FACT *Add percentage errors to get percentage error for a division.*

If you meet the square of a quantity, for example, V^2 in the formula, $P = V^2 / R$ then it must be remembered that $P = V \times V/R$ and the percentage error in each V must be combined (twice the percentage error in V) to get the percentage error in V^2. This is then added to the percentage error in R (a dividing quantity). Thus if $V = 12$ V, $R = 20\ \Omega$ and both have ± 1% possible errors then the possible error for P is ±2% for V^2 and ±1% for R giving ± 3% for P. Hence

$$P = \frac{12^2}{20} = 7.2\ \text{W} \pm 3\% = 7.2\ \text{W} \pm 0.2\ \text{W or } 7.2 \pm 0.2\ \text{W}.$$

Similarly the percentage error in a cubed quantity is 3 times and in a square root half of that in the quantity itself. The percentage error in a reciprocal, say $1/x$, is the same as for the x.

Possible errors will also combine if a calculation involves additions or subtractions.

Mass of liquid = (mass of liquid + beaker) − (mass of beaker)

= $(80 \pm 0.5$ g$) - (50 \pm 0.5$ g$)$

is an example. The highest possible value for the mass of liquid is 80.5 − 49.5 which is 31 g and the smallest possible is 29 g. For the 80 − 50 = 30 g the possible error is ±1 g. The absolute error sizes of 0.5 and 0.5 are added. The rule is explained in Appendix I.

KEY FACT *Add the **absolute** errors of the quantities involved to get the **absolute** error in the sum or difference.*

Error estimation can be tricky when a quantity appears more than once in a formula. Just to illustrate this consider x/x where x has a possible error of say ±2%. Adding the percentage errors for a division gives ±4% but x/x must equal precisely 1 with zero possible error. The problem with our rule lies in our considering + or − possibilities for the error in each x whereas for the same quantities both errors must have the same sign. Consequently the subtraction of percentage errors for the division will cause the errors to cancel to zero.

It would not be advisable to calculate possible error in R using the formula $R = \dfrac{R_1 \times R_2}{R_1 + R_2}$ where R_1 and R_2 both appear twice in the formula. The worked example on page 92 shows how this problem is overcome.

Example

(a) Calculate the heat given out by a 25 watt electric heater in 40 seconds.

(b) How accurate is your answer likely to be for the data given?

Answer

(a) The formula here is $Q = P \times t$ (heat produced = power × time).

$$\therefore \quad Q = 25 \times 40 = 1000 \text{ joule.}$$

(b) 25 watt implies a possible error of ±0.5 W which is a percentage error of

$$100 \times \pm \frac{0.5}{25}\% = \pm 2\%. \quad 40 \text{ s implies a possible error of } \pm 0.5 \text{ s or } 100 \times \pm \frac{0.5}{40}\% = 1.25\%.$$

The percentage possible error for the heat produced is ±2% plus ±1.25% which is ±3.25%.

For the absolute value of the possible error we calculate Q using the formula above to get

$Q = 1000$ joule. Then the ±3.25% means an absolute error of $\pm \dfrac{1000\,J}{100} \times 3.25 = \pm 32.5 \text{ J}$.

Example

Calculate the resistance for which a potential difference of 24 V gives a current of 2.5 A and using your answer calculate a value for the possible error suggested by the given data.

Answer

Using $R = \dfrac{V}{I}$ to calculate the resistance gives $R = \dfrac{24}{2.5} = 9.6\ \Omega$.

For 24 V the possible error is ±0.5 V and the percentage error is ±2.083%.

For 2.5 A the possible error is 0.05 A. a percentage error of ±2%.

The percentage errors *add* for a division (*V divided by I*) and so give ± 2.083 + ±2 or ± 4.083%.

Stating this error to the nearest whole percentage as suggested earlier gives the answer as ±4%.

Test Yourself

1 A bone has a circular cross-section with a diameter of 9 mm ± 10%. Calculate (a) the area of this cross-section and (b) the percentage possible error in this calculated area.

> HINT
>
> *(a) Area = $\pi \times radius^2$ or $\dfrac{\pi}{4} \times diameter^2$ (b) % Error in $diameter^2$ = 2 × % error in diameter.*

2 A balance recorded 30.00 g for a watch-glass and 72.54 g when some powder was placed on it. Calculate (a) the mass of the powder and (b) the absolute possible error in this mass.

> HINT
>
> *(a) Subtract (b) Add absolute errors.*

3 If $x = 5.0 \pm 2\%$ what is the percentage error (a) in x^2 (b) in $\sqrt{x}$?

7.4 Allowing for errors when calculating

In an examination it is not necessary to calculate and state the absolute or percentage error in your result unless you are specifically asked to do so. What definitely should be avoided is giving answers that imply a greater accuracy than is justified. The simplest rule is to give your answer with the same accuracy as the least accurate of the data used in the calculation. This precaution should be all that is necessary for most calculations. (By *data* we mean the figures given in the question.)

KEY FACT *Answers should have no greater accuracy than the poorest of the data used.*

If you want to work out the effect of combining the errors in your calculations you may find that with practice it will be quite quick to do so but time available is usually short.

Using significant figures as an estimate of possible error in data

A measurement of 99 is assumed to have a possible error of ± a half which means about ± 0.5%. However for 99.1 we have 0.05 in 99.1 which is about 0.05. For 999 we have ± half in 999, or 0.5 in 1000 approximately or 0.05 in 100 or about ±0.05%.

KEY FACT *Percentage error is smaller in values with more significant figures.*

For the same number of significant figures

KEY FACT *A greater measurement has a smaller percentage error.*

The quickest decision on the number of significant figures needed for an answer is:

KEY FACT *Find the smallest number of significant figures used in the data and give your answer to the same number of significant figures.*

This should be your favourite rule. It agrees with the principle that the answer should have about the same accuracy as the poorest of the data used. This advice is used in the following worked example.

Example
Use the formula $w = \lambda d/a$ to calculate the fringe spacing (or width) w expected in a Young's slits experiment with data as follows:

Slit separation, $a = 1.2$ mm
Distance, d from slits to the plane of the fringes $= 25.2$ cm
Wavelength, $\lambda = 632.8$ nm

Answer

$$w = \frac{\lambda \times d}{a} \text{ and } d = 0.252 \text{ m}, \lambda = 632.8 \times 10^{-9} \text{ m}, a = 1.2 \times 10^{-3} \text{ m}$$

$$\therefore \quad w = \frac{632.8 \times 10^{-9} \times 0.252}{1.2 \times 10^{-3}} = 1.329 \times 10^{-4} \text{ m} = 0.13 \text{ mm}.$$

As regards accuracy of the data used, the value of a is given to only 2 significant figures compared with 3 for d and 4 for λ. The answer therefore is reduced to 2 significant figures the same as for a, and is given as 1.3×10^{-4} m or 0.13 mm.

Rounding off

Removing some of the figures (or digits) from the right-hand side of a number is usually called **rounding off**. Consider a number say 23.4365 which is to be reduced to 3 significant figures because of the possible error in the number. We write 23.4 which is smaller than 23.4365. The number has been **rounded down**. In the case of reducing a number like 23.492 to 3 significant figures *rounding up* to 23.5 is chosen in preference to 23.4 because 23.492 is closer to 23.5. For rounding off a number such as 23.45, exactly half way between 23.4 and 23.5, rounding up is used.

A flow of 10 grams in 30 seconds means a flow rate of 10/30 or $\frac{1}{3}$ g per s and on your calculator the result may be shown as 0.3333333. The data are given only to 2 significant figures accuracy so the flow rate answer is given to 2 significant figures also and is rounded down to 0.33 g per s (0.33 g s^{-1}).

Rounding off should be left until the answer is obtained. However, during a calculation it is not convenient to write down and work with numbers with many figures. So you will mostly write numbers with one or two more significant figures than will be used for your answer and then round off the answer obtained to the appropriate number of significant figures.

> **KEY FACT**
> - *Decide the number of significant figures for the answer.*
> - *Use one or two more significant figures during calculation.*
> - *Round off answer to required number of significant figures.*

For a calculation in which one quantity is calculated as an answer and is then used to determine a second quantity the first answer is rounded off as necessary but its value with more significant figures is used for the subsequent calculation.

Some aspects of rounding off are illustrated in the following worked example.

Example
Calculate the work done in slowly raising a mass of 2.5 kg through a height of 1.35 m. (Acceleration due to gravity $g = 9.81$ m s^{-2}.)

Answer
The formula for the work done is $W = mgh$ so that $W = 2.5 \times 9.81 \times 1.35$

$$\therefore \quad W = 3.311 \times 10^1 \text{ joule.}$$

But the data used for the calculation includes a two figure number, namely the 2.5, so the answer cannot have greater accuracy than this. Rounding off the 3.311 to two figures gives 3.3.

So the work done is 3.3×10^1 or 33 joule.

Test Yourself

Exercise 7.4.1

1 Round off each of the following numbers to 2 significant figures:

   (a) 3.406   (b) 3.478   (c) 3.99×10^5

2 Calculate the volume in cubic metres of a rectangular tank measuring 0.60 m by 0.29 m by 0.29 m and give your answer to an appropriate number of significant figures.

Subtracting similar numbers

A calculation may require a number to be subtracted from a number that is not much bigger. The resulting difference could be smaller than the possible error for this subtraction and then the answer could be quite useless. In other cases the possible error in the difference has a less serious effect but is particularly important because of the smallness of the difference, the percentage possible error in the difference being large. The following worked example illustrates this problem.

> **KEY FACT** *Subtracting nearly equal numbers can result in an answer with a very high percentage error. Beware!*

You are not expected to calculate or estimate the possible error in such cases. Avoid such a situation when making measurements.

Example

A resistance of unknown value x kilohm (or kΩ) is placed in parallel with a resistance known to be 0.49 kΩ and the resulting combination then has a resistance of 0.47 kΩ. Calculate x.

Answer

> **HINTS AND TIPS** *Since the unit for all the resistances is kΩ the 10^3 that would be used for working in ohms will cancel when it divides every one of the three terms in the equation. It is therefore omitted. Try the calculation with it in if you wish.*

The equation relating the resistance R of two resistances R_1 and R_2 in parallel is $\dfrac{1}{R} = \dfrac{1}{R_1} + \dfrac{1}{R_2}$ and this can be rearranged so that $\dfrac{1}{R_1} = \dfrac{1}{R} - \dfrac{1}{R_2}$.

Replacing R by 0.47 and R_2 by 0.49 gives

$$\frac{1}{R_1} = \frac{1}{0.47} - \frac{1}{0.49} = 2.128 - 2.041 = 0.086\,66$$
$$\therefore \quad R_1 = 11.539 \text{ or } 11.539 \text{ k}\Omega.$$

Now the data used had 2 significant figure accuracy but can the answer be given to 2 significant figures accuracy?

The percentage error in the 0.49 kΩ value is $\pm$ half in 49 or about $\pm 1\%$ and for $\dfrac{1}{0.49}$ it is the same, i.e. about $\pm 1\%$, and the absolute possible error will be $\pm \frac{1}{100} \times 2.041 = \pm 0.02041$. For the $\frac{1}{0.47}$ there is a similar error of about ± 0.02 and for the difference the absolute error is about $\pm 0.02 + \pm 0.02 = \pm 0.04$ and this, as a percentage of 0.086 66, is almost 50%!! This figure also applies to R_1. Consequently the result has very poor accuracy and 2 significant figures are not justified for the answer.

Exam Questions

Exam type questions to test understanding of Chapter 7

Exercise 7.4.2

1 A leaf has a length of 87 ± 1 mm, a width of 44 ± 1 mm and a thickness of 0.2 ± 0.01 mm. What is the percentage possible error for each of these measurements?

2 Two resistors R_1 and R_2 each having a resistance of 1000 ohm $\pm 1\%$ are connected in series. Calculate the resistance R of this series combination and the percentage possible error for R.

HINT> *The formula required is $R = R_1 + R_2$.*

3 A volume of 100 ± 0.5 cm^3 of a certain solution weighs 112 ± 0.2 gram. Calculate (a) its density and (b) the percentage possible error for this density.

HINT> *(a) The formula required is density $= \dfrac{mass}{volume}$. (b) Add percentage errors.*

Answers to Test Yourself Questions

Exercise 7.1.1
1 (a) 34.35, 34.25 cm (b) 34.375, 34.365 cm
  (c) 9.95, 9.85 mm (d) 2.05, 1.95 m
  (e) 50.995, 50.985 mm
2 (a) 1.7 A (b) 2.2 A (c) 1.5 (d) 1.3 A
  (e) 2.0 A (f) 2.0 A (g) 4.0 A (h) 10.0 A

Exercise 7.1.2
1 (a) $2.5 \pm 2\%$ (b) $300 \pm 10\%$ (c) $6.25 \pm 0.08\%$
2 (a) 35 ± 3.5 (b) 72 ± 0.72 (c) 4.0 ± 0.05
  (d) 90 ± 0.27
3 C

Exercise 7.2.1
1 (a) 3 (b) 3 (c) 4 (d) 4 (e) 3
2 (a) ± 0.5 cm (b) ± 0.005 (c) ± 0.0005

Exercise 7.2.2
1 (a) $\pm 0.005 \times 10^3$ (b) 0.05×10^4 (c) 0.0005×10^3
  (d) 0.05×10^6

Exercise 7.3.1
1 (a) 63.62 mm^2 (b) 20%
2 (a) 42.54 g (b) ± 0.01 g
3 (a) 4% (b) 1%

Exercise 7.4.1
1 (a) 3.4 (b) 3.5 (c) 4.0×10^5
2 0.050 m^3

Chapter 8

Algebra, equations and transposition

After completing this chapter you should:

- *know how to rearrange an equation to obtain a useful formula*
- *have learnt more about presenting a calculation*
- *be able to simplify equations*
- *be able to calculate using proportionalities.*

8.1 Algebra – its meaning

What everyone seems to know about *algebra* is that it is a part of mathematics in which letters, mostly of the English alphabet, are used to represent numbers and measurements.

The well-known formula $T = 2\pi \sqrt{\dfrac{L}{g}}$ uses T for the period and L for the length of the pendulum and g for the gravitational intensity. We use symbols in this way when we want to save a lot of writing and when numerical values to replace them are not yet known.

Algebra also includes the rules that are needed when using these letters. You will for example meet the equation $v^2 = u^2 + 2as$. A level exam papers often contain lists of data and formulae and this equation may be found in such a list. But you may want a formula not for v^2 but for a. Then you will need to use some of the rules of algebra.

Letters (or *symbols*) can also be used for values that are too large or otherwise too inconvenient to be written repeatedly.

In maths classes and textbooks the most used letters are x and y. In physics, although the value represented by a letter is usually unknown, the name of the quantity (force or length, etc.) is known and a letter can be chosen that makes clear which quantity is represented (F for force, L for a length).

HINTS AND TIPS >   *Choose helpful symbols.*

There are no rules regarding whether small (lower case) or capital (upper case) letters should be used. But some quantities seem to have acquired particular letters, A is usually used for area and a for acceleration.

Because of the limited number of letters in the English alphabet some *Greek letters* are commonly used. π (*pi*) you have met but it will be explained more in Chapter 11 where θ (*theta*) and ϕ (*phi* pronounced fie) are also used. *Alpha* (α), *beta* (β) and *gamma* (γ) are often chosen for angles as well as being used to distinguish the three common nuclear radiations.

8.2 Equations, formulae and identities

Equations are statements that include equals signs and **formulae** are equations arranged so that the left-hand side names a quantity to be evaluated and the right-hand side is arranged so that it reads like a cooking recipe and shows how the required quantity is to be worked out.

The equation that is usually learnt for resistances in parallel is $\frac{1}{R} = \frac{1}{R_1} + \frac{1}{R_2}$. From this we get the *formula* for R which is $R = \frac{R_1 R_2}{R_1 + R_2}$.

Identities resemble equations but use $\equiv$ signs. The statement $3 + 5 = 8$ could be written as $3 + 5 \equiv 8$ because it is true under all circumstances. In contrast the statement $x^2 + 5x = 6$ is an equation that is true only if x equals 1 or -6. Identities were met in Chapter 5.

Rearranging equations

$x = 7$ is an equation that cannot be simplified or rearranged. The equation $\frac{6}{8 + R} = 0.5$ is less simple. This equation can be used to find the value of the unknown quantity, the resistance R, i.e. it can be *solved* for R. The equation can be made into a formula for R, namely

$R = R = \frac{6}{0.5} - 8$. The *rearranging* used is called **transposition**.

This formula shows that $R = 12 - 8$ or 4 ohm but how is the transposing of the equation done?

You may have realised that R was 4 ohm just by looking at the original equation but with less simple figures you would need rearrangement to provide the required formula.

The equation $\frac{6}{8 + R} = 0.5$ will be a useful example to rearrange but we start with a simpler example: $\frac{x}{5} = 3$.

Since $\frac{x}{5}$ equals 3 then 5 times $\frac{x}{5}$ will be the same as 5 times 3, i.e. $5 \times \frac{x}{5} = 5 \times 3$
Then cancelling the fives on the left gives $x = 5 \times 3 = 15$.

Why was multiplying by 5 chosen?

Because it removed the 5 beneath the x.

The equation $x \times 3 = 6$ can be simplified by dividing both sides of the equation by 3 to remove the 3 from the left and leave the equation $x = \frac{6}{3}$.

Another simple example is $x + 3 = 5$ and we would prefer to have $x = something$. To remove the 3 from the left we can say that since $x + 3$ has the same value as 5 then 3 less than the $x + 3$ will equal 3 less than 5 and so $x + 3 - 3 = 5 - 3$. The $+ 3$ and $- 3$ cancel to give $x = 5 - 3$.

In a similar way $x - 4 = 9$ can be simplified by adding 4 to each side of the equation.

So useful steps in rearranging an equation can be:

Multiplying or dividing both sides by a chosen number or adding or subtracting a chosen number from both sides.

The steps above can be summarised in the following two rules:

RULE 1 *A number dividing everything on one side of an equation can be moved to the other side where it must multiply everything else there. For x/5 = 3 the 5 dividing on the left is replaced by 5 multiplying on the right. So x = 3 × 5.*

Reading the rule in reverse, a number multiplying everything on one side can be moved to the other side where it must divide everything else there.

RULE 2 *A number adding to everything on one side of an equation can be moved to the other side where it must subtract from everything else.*

Using this rule $x + 3 = 5$ becomes $x = 5 - 3$.

Similarly for the equation $x - 4 = 9$ the $- 4$ on the left becomes $+ 4$ on the right to give $x = 9 + 4$.

The words *everything else* in these rules are important. In the case of $\dfrac{x+2}{3} = 4$ which requires x to be 10, it would be incorrect to say that the 2 is adding on the left and so will be subtracted on the right. This would give $\frac{x}{3} = 4 - 2 = 2$ (Oh, dear!). The 2 was *not* adding to *everything else* on the left. It was not adding to the 3. It would have been adding to everything else if the equation had been $\frac{x}{3} + 2 = 4$.

So in the $\dfrac{x+2}{3} = 4$ equation the first move will have to use rule 1 and change the 3 from dividing everything else on the left (the $x + 2$) to multiplying everything else on the right (the 4).

$$\frac{x+2}{3} = 4 \text{ becomes } x + 2 = 4 \times 3 \qquad \therefore \ x + 2 = 12$$

The new equation has 2 adding to everything on the left and this 2 can now be moved to the right (rule 2) where it will subtract from everything else there (the 12).

$$x + 2 = 12 \text{ becomes } x = 12 - 2 = 10$$

The aim when rearranging an equation is to get the unknown quantity (say x) on its own so that the equation has the form

$$x = \dots \text{ or } \dots = x$$

It is usual to try to get the x on the left because we read from left to right and we try to get it there on its own.

If there is a denominator containing the unknown quantity this can be a nuisance and we usually try to remove the unknown from any denominator. This approach is followed in the following worked example.

Example

$$\frac{6}{8+R} = 0.5$$
$$\therefore \quad 6 = 0.5 \times (8 + R)$$
$$\therefore \quad \frac{6}{0.5} = 8 + R$$
$$\therefore \quad 12 = 8 + R$$
$$\therefore \quad 12 - 8 = R$$
$$\therefore \quad 4 = R$$
$$\therefore \quad R = 4 \text{ ohm}$$

The symbol $\therefore$ means *therefore* or *it follows that* or *this leads to*.

In the first step of the above calculation the $8 + R$ as a whole is moved because the R on its own was not dividing everything else on the left. Once on the right of the equation the $8 + R$ was enclosed in brackets because the whole of this $8 + R$ must multiply the 0.5, or, the 0.5 must multiply the whole of the $8 + R$.

The 0.5 was then moved to the left of the equation using the *multiplying becomes dividing* rule. Neither the 8 nor the R on the right could be moved at this stage because neither was multiplying everything else or adding to everything else on that side. Subsequently the 8 could be moved to leave R on its own.

Examples

1 $\dfrac{2x + 4}{9} = x - 2$, determine x.

Answer

The 2 on the right could be moved to the left of the equation but then no further move could be made. So we return to the start and try again. The 9 can be moved to give $2x + 4 = (x - 2) \times 9$ and removing the brackets gives $2x + 4 = 9x - 18$.

We should aim at getting all x terms together on one side of the equation with all numbers on the other side. Moving the 4 gives $2x = 9x - 18 - 4$ or $2x = 9x - 22$.

Now the $9x$ is moved and the equation becomes $2x - 9x = -22$ so that $-7x = -22$. This means that the $7x$ is equal to 22 or $7 \times x = 22$.

Finally the 7 is moved and the answer is $x = 22/7 = 3.14$.

2 $1 - \dfrac{5x}{4} = 3$ can be changed as follows:

$$\therefore \quad 1 - \frac{5x}{4} = 3$$

$$\therefore \quad \frac{-5x}{4} = 3 - 1 = 2$$

$$\therefore \quad -5x = 2 \times 4$$

$$\therefore \quad -x = \frac{2 \times 4}{5} = \frac{8}{5}$$

but we want $x = \dots$ not $-x = \dots$ so view the equation as $-1 \times x = 1.6$ then $x = \frac{1.6}{-1}$ and multiplying top and bottom of this fraction by -1 gives -1.6 for x.

KEY FACT *If $-x = 1.6$ then $x = -1.6$.*

Example

In a certain calorimetry calculation the following equation was obtained:

$$4.2 \times 10^3 = 420 \times (x - 15)$$

Calculate the temperature, x.

Answer

We need to transpose the equation to get $x = \dots$

Using rule 1 above gives

$$\frac{4.2 \times 10^3}{420} = x - 15 \text{ or } 10 = x - 15$$

The *symbol* ∝ is used for *is proportional to* so that you can meet potential difference ∝ current.

KEY FACT   *x* ∝ *y* means *x* is proportional to *y*

*Fig. 1*a shows an electric circuit in which the current can be measured in amperes for each potential difference (p.d.) provided by the voltage supply. The results are shown in the table (*Fig. 1b*).

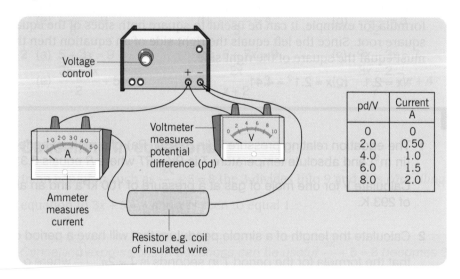

pd/V	Current A
0	0
2.0	0.50
4.0	1.0
6.0	1.5
8.0	2.0

Fig. 1 *Electric circuit for current and potential difference measurements.*

When the potential difference is changed from 2.0 V to 4.0 V the current doubles from 0.5 A to 1.0 A and for the change from 2.0 V to 8.0 V the current is quadrupled to 2.0 A. So the current and potential difference are proportional to each other.

Graphs and proportionality

If a graph is drawn of the electrical measurements just mentioned and listed in *Fig. 1*, the graph is as shown in *Fig. 2*.

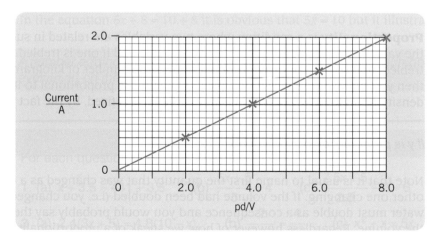

Fig. 2 *Graph showing proportionality.*

It is a *straight-line graph*. The graph line is a straight line and passes through the point where the current is zero and the potential difference is zero. This means that the current and potential difference are *proportional*.

KEY FACT *For a graph to show proportionality it must be a straight line that passes through the point (0, 0).*

Fig. 3 shows a graph in which x and y are not proportional. Changing x from 2 to 4 does *not* double y from 2.5 to 5 but takes y from 2.5 only up to 3.

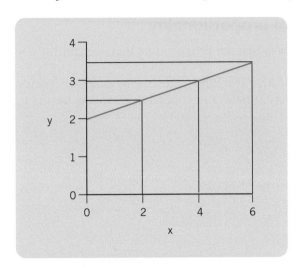

Fig. 3 *Variables NOT proportional*

However any *change* in x causes a proportional *change* in y. For example if x changes from 2 to 4 it causes a change in y from 2.5 to 3 – a change of 0.5. If now a change in x twice the previous size is considered, e.g. from 2 to 6 then the y change is from 2.5 to 3.5 – a change of 1 which is twice what it was before.

KEY FACT *For any straight-line graph changes in y are proportional to changes in x.*

An equation for proportionality

If $y \propto x$ then we can always write an equation $y = kx$ in which k stands for a **constant**. By a *constant* we mean a quantity the value of which is unaffected by changes in x and y. It remains the same during these changes.

Using I for current and V for potential difference for the results shown in *Fig. 1b* we expect $I = kV$ to apply.

The result $V = 2$ volt when $I = 0.5$ A gives $k = 0.25$. Its unit is ampere/volt.

If V is changed to 6 for example then the equation predicts that $I = 0.25 \times 6 = 1.5$ A and this is correct.

KEY FACT *If x and y are proportional then x = ky where k is a constant or y = kx where k is a different constant.*

k is often called the **constant of proportionality**.

The value of k found above, namely 0.25 (unit ampere per volt) applies to $I = kV$ or $k = I/V$. If instead the relationship is written as $V = kI$ then $k = V/I$ and is the reciprocal of I/V, its previous value. This is illustrated by again using the result $V = 2.0$ V when $I = 0.5$ A which gives $k = 2.0/0.5$ which equals 4. This 4 is the reciprocal of the previous 0.25.

Now that V/I has been mentioned it is relevant to point out that V/I for a resistor like the one in *Fig. 1(a)* is used as the measure of its resistance (R) $R = V/I$. So the resistance there has a value of 4 and the *volt per ampere* unit for R is called the ohm (Ω).

A consequence of proportionality

If x and y are two variables that are proportional then we can say $y = kx$.

Now when x is first at a value x_1 let the value of y be called y_1. Then $y_1 = k \times x_1$. If then x changes to a new value x_2 causing y to change to a value y_2 we get $y_2 = k \times x_2$. Consequently $y \propto x$ means

$$\frac{y_2}{y_1} = \frac{k \times x_2}{k \times x_1} = \frac{x_2}{x_1}$$

KEY FACT $y \propto x$ means $\dfrac{y_2}{y_1} = \dfrac{x_2}{x_1}$.

Example
A fixed mass of a perfect gas contained in a fixed volume is at a pressure of 100 kPa and its absolute (kelvin) temperature is 273 K. What is its expected pressure if the temperature were to be raised to 373 K, given that the pressure is proportional to the absolute temperature?

Answer

Let $P = P_1$ at initial temperature T_1 and $P = P_2$ at T_2. Then $\dfrac{P_2}{P_1} = \dfrac{T_2}{T_1}$

Using pascal (Pa) as the unit for P, $P_1 = 100 \times 10^3$ Pa so that $\dfrac{P_2}{100 \times 10^3} = \dfrac{373}{273} = 1.37$

$$\therefore \quad P_2 = 1.37 \times 100 \times 10^3 = 1.37 \times 10^5 \text{ Pa}$$

If you could clearly see in advance that the 10^3 used to convert to pascals would cancel when division by 10^3 converts the answer to kilopascal then these conversion figures could be omitted.

Test Yourself

Exercise 8.3.1

1 A mass m on a spring has a period of 1.2 second what will the period be if the mass is replaced by one 4 times greater using the same spring? The formula for the period is

$T = 2\pi \sqrt{\dfrac{m}{K_s}}$ where m is the mass and k_s is the *spring constant* (a characteristic of the spring).

HINT

$$\frac{T_2}{T_1} = \frac{\sqrt{m_2}}{\sqrt{m_1}} = \sqrt{\frac{m_2}{m_1}}$$

2 The potential difference (V) between the two ends of a resistor is 3.0 volt and produces a current (I) of 6.0 mA. If this voltage is increased to 12 volt what current will flow if the resistance of the resistor remains constant so that I and V are proportional?

HINT $\dfrac{V_2}{V_1} = \dfrac{I_2}{I_1}$

Proportion and ratio

If one quantity is 6 times bigger than a second quantity we say their ratio is 6 to 1 or 6/1. So a ratio is one quantity divided by another, used to compare the quantities For example the ratio of 1 hour to 1 minute is

$$\frac{1 \text{ hour}}{1 \text{ minute}} = \frac{60 \text{ minute}}{1 \text{ minute}} = 60$$

KEY FACT *The ratio of y to x is $\frac{y}{x}$.*

The ratio of two variables such as mass and volume can only be constant if these variables are proportional to each other because then doubling the numerator of the ratio doubles the denominator also and the ratio stays the same.

KEY FACT *Proportionality means constant ratio.*

Sharing *in proportion to* is met for example when a voltage is shared (or divided) between two resistances in proportion to the resistance values. So 10 volts shared between a 3 ohm resistance and a 2 ohm resistance produces voltages V_3 and V_2 such that $\dfrac{V_3}{V_2} = \dfrac{3}{2}$.

Now

$$10 = V_3 + V_2 = \frac{3V_2}{2} + V_2$$

$$\therefore \; V_2 = \frac{10 \times 2}{5} = 4 \text{ volts, leaving 6 volts for } V_3.$$

You can get this result easily by dividing the 10 volts into 5 portions (5 because 3 + 2 is 5). You get 2 volts for a portion. Then 3 portions (i.e. 6 volts) are allocated to the 3 ohm resistance and two portions (4 volts) to the 2 ohms.

Example
If the volume of a certain gas under given conditions is in constant ratio with its absolute temperature how will the volume be affected by doubling the absolute temperature?

Answer
$$\frac{V_2}{V_1} = \frac{T_2}{T_1} = 2$$
$\therefore \; V_2 = 2 \times V_1$ so the volume is doubled.

Inverse proportionality

Inverse here means upside down and y is *inversely proportional* to x if $y \propto 1/x$ and the corresponding equation is $y = k \times \dfrac{1}{x}$ or $y = \dfrac{k}{x}$ or $x \times y = k$ or $xy = k$ or xy is constant

KEY FACT *Inverse proportionality, $y \propto \frac{1}{x}$.*

For this condition $\dfrac{y_2}{y_1} = \dfrac{k/x_2}{k/x_1} = \dfrac{k \times x_1}{k \times x_2} = \dfrac{x_1}{x_2}$

KEY FACT *For inverse proportionality $\dfrac{y_2}{y_1} = \dfrac{x_1}{x_2}$.*

To distinguish between inverse proportionality ($y \propto \frac{1}{x}$) and the proportionality $y \propto x$ the term **direct proportionality** is used when necessary for the $y \propto x$ condition.

Note that that the $\dfrac{x_2}{x_1}$ equated to $\dfrac{y_2}{y_1}$ for direct proportionality is inverted to $\dfrac{x_1}{x_2}$ for inverse proportionality.

Example

1 The volume V, of a fixed mass of gas at constant temperature is inversely proportional to the pressure P, of the gas. If the initial volume of such a gas sample is V_1 at pressure P_1 what will its new volume be when the pressure is 7 times greater?

Answer

$$\frac{V_2}{V_1} = \frac{P_1}{P_2} = \frac{P_1}{7 \times P_1} = \frac{1}{7} \qquad \therefore \ V_2 = V_1/7$$

2 The illumination of a surface (the *illuminance*) caused by a small lamp is inversely proportional to the square of the distance of the lamp from the surface (*see Fig. 4*). Hence the illuminance (I) is proportional to $1/d^2$ or $I = k/d^2$. What is the effect on I of doubling the distance?

Answer

$$\frac{I_2}{I_1} = \frac{1/d_2^2}{1/d_1^2} = \frac{d_1^2}{d_2^2} = \frac{d_1^2}{(2 \times d_1)^2} = \frac{d_1^2}{4 \times d_1^2} = \frac{1}{4}$$

So the new illuminance is a quarter of the original value.

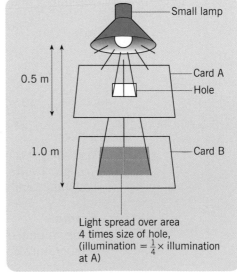

Fig. 4 Illumination of a surface.

Test Yourself

1 The lens power (P) of a biconvex lens is inversely proportional to the radius of curvature (R) of its surfaces. One biconvex lens has R equal to 15 cm and its power is P_1 and a second biconvex lens made of the same type of glass has R equal to 10 cm and its power is P_2. What is the ratio of P_1/P_2?

HINT

$$\frac{P_2}{P_1} = \frac{R_1}{R_2}$$

2 The force (F), between two small electric charges 10 cm apart in air is 1.2×10^{-6} N. What will the size of this force be if the distance (d), between the charges is increased to 12 cm, given that F is inversely proportional to d^2?

HINT

$$\frac{F_2}{F_1} = \frac{d_1^2}{d_2^2}$$

3 The resonant frequency (f), for a series combination of a fixed inductance and a capacitance (C), is inversely proportional to the square root of C. If C is changed from 1.00 μF to 1.44 μF by what factor is f multiplied?

Examples

Question 1 below is different from previous examples because more than one variable changes and affects the answer.

1 A mass hanging on a spring is disturbed so that it oscillates with a period of 0.6 s. The formula for the period is $T = 2\pi \sqrt{\dfrac{m}{k}}$. What will the new period be if the mass were four and a half times bigger and the spring is replaced by one with a spring constant k twice that of the first spring?

Answer

Let the first period be T_1, first mass m_1 and first k be k_1 and use subscript 2 for the new values.

$$T_1 = 2\pi \sqrt{\frac{m_1}{k_1}} \quad \text{and} \quad T_2 = 2\pi \sqrt{\frac{m_2}{k_2}} \quad \text{so that} \quad \frac{T_2}{T_1} = \frac{2\pi\sqrt{m_2/k_2}}{2\pi\sqrt{m_1/k_1}}$$

and squaring both sides of the last equation to remove the square roots gives $\left(\dfrac{T_2}{T_1}\right)^2 = \dfrac{m_2/k_2}{m_1/k_1}$ and by multiplying top and bottom of the right-hand side by k_1 and k_2,

$$\left(\frac{T_2}{T_1}\right)^2 = \frac{m_2 \times k_1}{m_1 \times k_2} = \frac{4.5 \times m_1 \times k_1}{m_1 \times 2 \times k_1} = \frac{4.5}{2} = 2.25$$

$$\therefore \quad \frac{T_2}{T_1} = \sqrt{2.25} = 1.5$$

$$\therefore \quad T_2 = 1.5 \times T_1 = 1.5 \times 0.6 = 0.9 \text{ s}$$

Note here that the 2π was really the constant of proportionality and cancelled.

The approach used here has been to write down the required ratio T_2/T_1 and then replace each of its two quantities (T_2 and T_1) by a formula.

> **KEY FACT** *A ratio can be found by writing it first as a ratio of two formulae.*

This approach is particularly reliable because all steps in the calculation are written down. You may prefer it when more than one quantity changes and affects the ratio or when the factors by which the quantities change are not simple values.

However, it can be said here that k is doubled and $\sqrt{k}$ is therefore multiplied by $\sqrt{2}$ which is 1.414. This change would decrease the period by the factor 1.414. Multiplying the mass by 4.5 would increase $\sqrt{m}$ by a factor of $\sqrt{4.5}$ which is 2.121. So the overall effect is to multiply the 0.6 s by 2.121 and divide by 1.414 and this gives 0.9 correctly. However this method needs a lot of thinking and involves too much risk of mistake. Also the previous method can be written down and understood more easily by the reader (teacher or examiner perhaps).

Example
In question 2 that follows the proportionality involves a power of R.

2 The period of a satellite moving round the earth in a circular orbit of radius R is T. If the orbit were changed to a radius $4R$ what would the new period be?

    A $\frac{1}{4}T$ **B** T **C** $4T$ **D** $8T$ **E** $64T$

Answer
Some syllabuses require Kepler's third law to be known. It says that $T \propto R^{\frac{3}{2}}$.

$$\therefore \quad \frac{\text{new } T}{\text{old } T} = \frac{k(4R)^{3/2}}{kR^{3/2}} = 4^{3/2} = 2^3 = 8 \text{ which is answer D.}$$

A briefer answer is obtained by saying $T^2 \propto R^3$ so that T^2 increases by a factor of 4^3 or 64 and T increases by a factor of $\sqrt{64}$ or 8.

Test Yourself

Exercise 8.3.3

1 The heat flow in watts (Q) through a certain metal rod of cross-section area A and length L with a temperature difference θ between its ends is given by the formula $Q = \dfrac{kA\theta}{L}$. The area $A = \pi R^2$ is a constant ($\pi = 3.142$) and R is the rod radius. k is the thermal conductivity of the metal concerned. If Q were measured for this rod and then for a rod having twice the radius and half the length and three times the temperature difference but made of the same metal how would the new Q compare with the old value?

HINT $\dfrac{Q_2}{Q_1} = \dfrac{R_2^2 L_1 \theta_2}{R_1^2 L_2 \theta_1}$

2 The force per unit length (F/L) on a straight wire carrying a current I and lying parallel to a long straight wire carrying an equal current is directly proportional to the square of the current and inversely proportional to the distance d between the wires. If d is changed from 4.0 cm to 3.0 cm and the current is doubled by what factor is F/L multiplied?

HINT *You could assume that $I_1 = 1$ A and $d_1 = 1$ m.*

Example

By what factor must the voltage across a coil of wire be increased if the coil's resistance increases by a factor of 1.2 and the current through the coil is to be doubled?

Answer

Voltage V = current I × resistance R so that $V \propto I$ and $V \propto R$.

The change of voltage can be considered in two stages, the first being a change associated with the current change and the second associated with the resistance increase.

For the first stage where $V \propto I$

$$\frac{\text{new } V}{\text{old } V} = \frac{\text{new } I}{\text{old } I} = 2 \qquad \therefore \quad \text{new } V = 2 \times \text{old } V$$

For the second stage where $V \propto R$

$$\frac{\text{final } N}{\text{new } V} = \frac{\text{new } R}{\text{old } R} = 1.2 \qquad \therefore \quad \text{final } V = 1.2 \times \text{new } V = 1.2 \times 2 \times \text{old } V = 2.4 \times \text{old } V.$$

The factor by which V is increased is 2.4.

KEY FACT   *The effect of changes by given factors can be calculated in stages.*

Exam Questions   **Exam type questions to test understanding of Chapter 8**

Exercise 8.3.4

1 It is suggested that the battery in a car could be replaced by a capacitor. A typical car starter motor draws a current of 150 A from the 12 V battery for 5 s to start the car.

Suppose a capacitor were to be used to store the *energy* for just one start. Which one of A to D below is the capacitance, in F, of the capacitor which would be required?

A 25   **B** 62   **C** 125   **D** 150

(OCR 2000)

HINT   *Energy = voltage × current × time and energy = $\frac{1}{2}C$ × voltage² where C is capacitance in F (farad).*

2 Ultraviolet light of wavelength 12.2 nm is shone on to a metal surface. The work function of the metal is 6.20 eV.

(a) Calculate the maximum kinetic energy of the emitted photoelectrons.
(b) Show that the maximum speed of these photoelectrons is approximately 6×10^6 m s^{-1}.
(c) Calculate the de Broglie wavelength of photoelectrons with this speed.
(d) Explain why these photoelectrons would be suitable for studying the crystal structure of a molecular compound.

(Edexcel 2001)

HINT   *$\frac{hc}{\lambda}$ = WFE + KE max. KE = $\frac{1}{2}mv^2$. De Broglie wavelength = $\frac{h}{p}$, h = 6.63 × 10⁻³⁴ J s, c = 3.00 × 10⁸ m s⁻¹, m = 9.11 × 10⁻³¹ kg, λ denotes the wavelength of light, v is velocity and p momentum (mv) of photoelectron. 1 nm = 10⁻⁹ m, 1 eV = 1.6 × 10⁻¹⁹ J.*

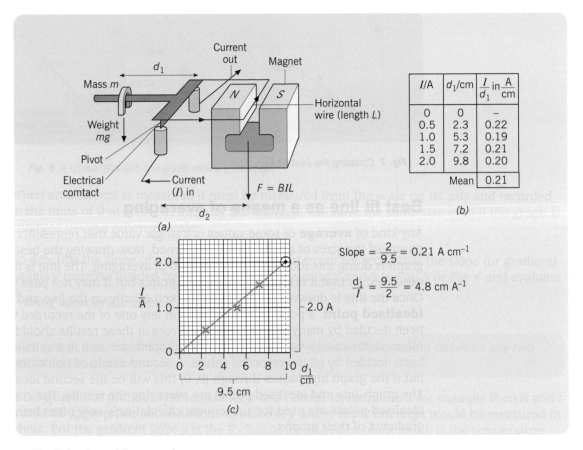

Fig. 8 *Graph providing averaging.*

The force F on the wire is given by the formula $F = BIL$ where B is the strength of the magnet's field, called the magnetic flux density, measured in teslas. The force F acting downwards tries to tilt the see-saw by producing a clockwise turning effect (or moment). $F \times d_2$ but the small mass m of 1.0 gram is moved to the necessary distance d_1 to produce an equal anticlockwise moment of $mg \times d_1$ and this keeps the see-saw horizontal. g is the acceleration due to gravity = 10 m s^{-2}, d_2 is 12 cm, L is 4.0 cm.

Various current values (I) are used and the corresponding d_1 value is recorded for each current. These are shown in *Fig.* 8b. Determine the flux density produced by the magnet.

Answer

The graph plotted is shown in part (c) of *Fig.* 8. A point has been marked by a dot, surrounded by a circle to emphasise it, placed exactly on the graph line. Its coordinates are $I = 2.0$ A and $d_1 = 9.5$ cm. These values for I and d_1 are idealised values used to calculate B as follows.

The moment due to F is $F \times d_2$ which equals $BILd_2$. The equal moment is mgd_1.

$$\therefore \quad mg\, d_1 = BILd_2 \text{ and } B = \frac{mgd_1}{ILd_2}$$

Using $I = 2.0$ A with $d_1 = 9.5$ cm gives $B = \dfrac{1 \times 10^{-3} \times 10 \times 9.5 \times 10^{-2}}{2 \times 4 \times 10^{-2} \times 12 \times 10^{-2}} = 0.10$ tesla.

This answer fed the coordinates of the idealised point into the calculation as 2.0 A for I and 9.5 cm for d_1. We usually say that the slope of the graph is $\frac{2}{9.5}$ ($= 0.21$ A cm^{-1} = 21 A m^{-1}) and use this for I/d_1 in the formula for B.

9.2 Equations for graphs

Equation for a graph showing proportionality

Graphs show relationships but equations showing relationships allow calculations and predictions to be made easily. Can we get an *equation* from a graph?

As explained in Chapter 8, if y is proportional to x then we can write $y = kx$ where k stands for a fixed quantity (a constant, one that does not change when x and y change).

Now in our graph showing *proportionality* (see *Fig. 4*) $\frac{y}{x}$ (for any point on the graph line) equals the slope of the graph or $\frac{a}{b}$ (measured for point P). It is usual to denote this constant by m in place of k, so that $m = \frac{y}{x}$ for any point on the line and $y = mx$. This slope or gradient m is found by measuring a and b for a point such as P. Thus $y = mx$ is the equation describing the straight-line passing through (0, 0). It uses m for the gradient although it is identical to the k we have used previously. So $y = mx$.

KEY FACT $y = mx$ for a graph showing proportionality and m is its gradient.

Test Yourself   Exercise 9.2.1

1 For the graph shown in *Fig. 1*(c) determine (a) the current value when the voltage is 3.0 volt and (b) the slope of the graph.

2 The table below records the charge (Q millicoulomb) stored on a capacitor when a variable voltage V is used. Plot a graph of Q on the y-axis versus V (x-axis) and measure the gradient.

Potential difference V/V	5.0	10	15	20	25	30
Charge Q/mC	0.25	0.51	0.77	0.99	1.24	1.50

An equation for a straight line not passing through point (0, 0)

Remembering that for a straight line passing through (0, 0) the equation is $y = mx$, we can now say that when instead there is an intercept on the y-axis of size c the equation is $y = mx + c$. This makes sense because the graph differs from the $y = mx$ graph only in that the whole line is raised by an amount c in the y direction. Every point has a y value increased by this amount c. If $y = 2x$ is compared with $y = 2x + 7$ the y values for $y = 2x$ when $x = 0, 1, 2, 3$ are 0, 2, 4, 6 and for $y = 2x + 7$ the y values are 7, 9, 11, 13, each y value being increased by 7, the value of the c term.

KEY FACT   For a straight-line graph with y intercept $y = mx + c$.

In the equation $y = mx + c$ the c is an appropriate symbol because it denotes a constant. It does not change when x and y change. It is the fixed amount added to every y value as you picture the line $y = mx$ being raised to a position with intercept c.

When describing a graph with a straight line and intercept we cannot say y is proportional to x but we can say that any *increase* in y is proportional to the corresponding *increase* in x.

KEY FACT   *For any linear relationship increase in y is proportional to increase in x.*

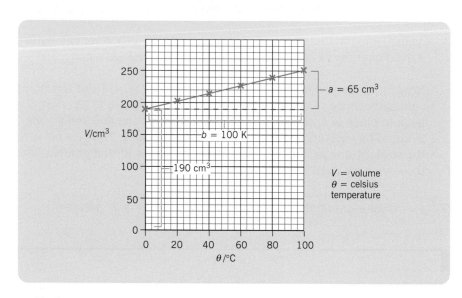

Fig. 9

Example
The graph in *Fig. 9* relates the volume of a gas kept at atmospheric pressure to its temperature which is varied. Obtain the equation relating the volume in cm^3 to its temperature θ.

Answer
From the graph $a = 65$ cm^3 and $b = 100$ K (0 to 100 °C).

Hence the gradient $= \frac{65}{100}$ cm^3 K^{-1}.

Remembering $y = mx + c$ for a straight-line graph, we have V for y, θ for x, $c = 190$ cm^3 and $m = 0.65$ cm^3 K^{-1} so that $V = 0.65\theta + 190$.

Test Yourself

Exercise 9.2.2

1 The graph in *Fig. 10* relates the pressure of some air kept at constant volume but a variable temperature. Obtain the equation relating the pressure P measured in cm of mercury to the temperature θ measured in °C or K.

HINT   *P = 95 when θ = 80 so slope is $\frac{95}{80}$ cm K^{-1}.*

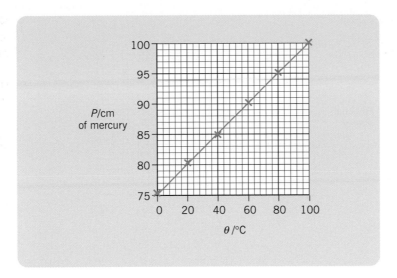

Fig. 10

2 The velocity v and time t for a certain vehicle movement were recorded as follows:

t/s	2	4	6	8	10	12
v/m s^{-1}	35	41	45	49	56	60

The equation for this movement is $v = u + at$. The u is the initial velocity at time $t = 0$ and a is the constant acceleration. Plot a graph of the results and hence deduce u and a.

Negative slopes

Not all graphs show a line that rises from left to right. The line could be horizontal or may fall. For a fall the gradient is negative. For example a graph relating the temperature of an object on the y-axis to time of cooling on the x-axis has a line that descends. The example below involves a graph whose gradient is not only negative but also is constant, this last feature being shown by the straightness of the graph-line.

KEY FACT *A graph that falls from left to right has a negative gradient (negative slope).*

Example
The graph in *Fig. 11* plots the gravitational potential energy (PE) of a falling object on the y-axis versus the vertical distance fallen through on the x-axis.

Answer
From the graph, b is the distance fallen, 20 m. The a is the consequent reduction of PE from 600 J to 200 J so that a is a *reduction* of 400 J, i.e. −400 J. The gradient of the graph (a/b) is then − 400 J/20 m which is −20 J m^{-1}.

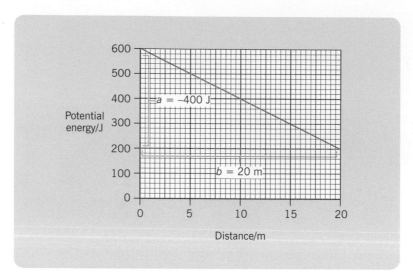

Fig. 11 *Graph relating to a falling body.*

Note that a positive gradient would have been obtained if the PE were plotted versus height (the distance measured upwards).

9.3 Graphs that are curves

Gradient of a curve

For a graph showing a *curve* the gradient is not constant, the steepness changes. In theory the gradient at a chosen point on the graph could be evaluated by measuring, in the vicinity of that point, the small change there in y (denoted by δy) that accompanies a small change δx in x, small enough that the curve behaves almost as a straight line over this small change. The slope is then $\frac{\delta y}{\delta x}$. The symbol δ is pronounced *delta* and stands for 'a small increase in'. It may also be denoted by a small triangle, Δ, (e.g. as in $\frac{\Delta y}{\Delta x}$).

In theory you could measure this with δx extremely close to zero and we have $\frac{\delta y}{\delta x}$, $\delta x \to 0$ which is written as $\frac{dy}{dx}$.

Fig. 12(a) illustrates the idea of measuring the gradient over a small length of curve but *Fig. 12(b)* shows the practical approach we need. It uses the fact that the slope of a *tangent* to a curve is the same as the slope of the curve where it touches.

In *Fig. 12(b)* points Q and R are marked equally to the left and right of P. The straight line through Q and R has a slope almost equal to that of the curve at P but only if Q and R are very close to P. As Q and R are brought nearer to P the straight line becomes a tangent to the curve. So as a practical method of measuring a slope a tangent is drawn to the curve and the slope of the tangent is measured as in *Fig. 12(c)*.

KEY FACT   *The slope of a curve at a point equals the slope of a tangent there.*

An advantage of this measurement, seen in *Fig. 12(c)*, is that a large value for b and therefore for a can be used so that accurate measurement is possible.

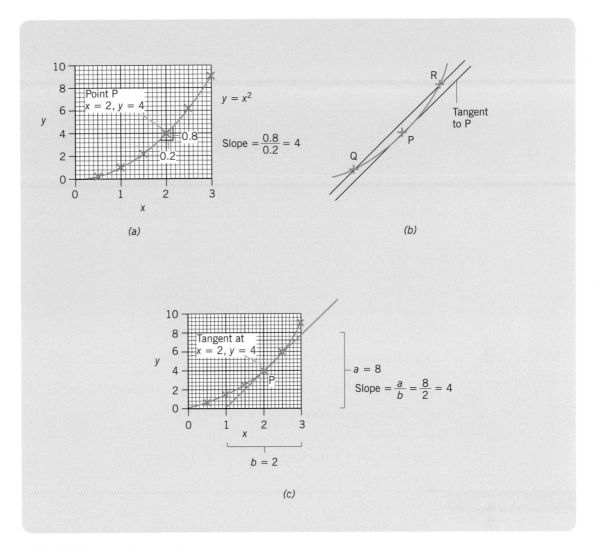

Fig. 12 *Measuring the slope of a curve.*

Example

The drain current of a field-effect transistor is measured for various values of the controlling gate-source voltage with the drain-source voltage kept constant.

Gate voltage/V	0.25	0.50	0.75	1.00	1.25	21.5	1.75	2.00
Drain current/mA	4.0	3.0	2.0	1.3	0.80	0.50	0.30	0.20

Plot the graph of drain current as ordinate versus gate voltage as abscissa and measure the slope of the graph for a gate voltage of (a) 0.5 V and (b) 1.5 V.

Answer

The graph is shown in *Fig. 13.*

The points on the graph where the slopes are to be measured have been marked by dots and small rings and a dashed line has been carefully drawn as a tangent at each of these points.

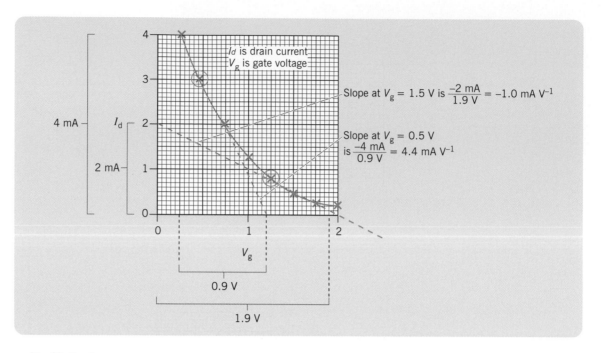

Fig. 13 *Graph.*

(a) For the point where $V_g = 0.5$ V the slope is a drop of 4 mA for a V_g change of 0.9 V. The slope is therefore $\dfrac{-4 \text{ mA}}{0.9 \text{ V}} = -4.4$ mA V^{-1}.

(b) At $V_g = 1.5$ V the slope is $\dfrac{-2 \text{ mA}}{1.9 \text{ V}} = -1.0$ mA V^{-1}.

Test Yourself

Exercise 9.3.1

1 Use the numbers given in the following table to produce a y versus x graph and measure the gradient at the point where $y = 20$.

x	0	1.0	2.0	3.0	3.5	4.0
y	1	2.7	7.4	20	33	55

2 The following results were obtained for a freely-falling object. Plot a graph of distance fallen as ordinate versus time as abscissa and measure the slope of the graph (a) at time 2 s and (b) at time 4 s.

Time/s	1.0	2.0	3.0	4.0	5.0
Distance/m	10	20	45	80	125

9.4 Other uses for graphs

Making predictions

A graph can sometimes be used to make predictions. Suppose that we want to predict the size of current that will flow when using a potential difference of 5.0 volt for the circuit shown in *Fig. 1*(*a*). The answer can be read from the graph of *Fig. 1*(*c*) and on the line at $V = 5.0$ V the current is 0.25 A.

Another possibility is that we may want to know the current to expect for a potential difference of 12 V, a voltage greater than any used to plot the graph. If we extend the line of the graph beyond the last plotted point we read 0.60 A for a potential difference of 12 V. A procedure like this which involves going outside of the range covered by the plotted points is called **extrapolation** whereas the result for 5.0 V was found from a place on the line and *between* plotted points so that it involves **interpolation** (*inter* meaning between and *extra* meaning outside or additional).

Extrapolation tends to be risky because you can't always be sure that the graph will behave as expected under circumstances not covered by the experiment that produced the graph. For example, a graph such as that shown in *Fig. 1*(*c*) will often cease to be straight when the current becomes large.

KEY FACT   *A graph can predict results by interpolation or by extrapolation.*

Discovering an equation that describes a graph

Often a graph is used to discover a relationship between two variables but at other times, and frequently in A level work, it is already known or supposed that the variables obey a particular equation and the purpose of the experiment is to confirm the equation and determine the value of a constant quantity in the equation. When we use a graph to do this we like to obtain a straight line. A curve is much more difficult to relate to an equation.

KEY FACT   *We like to get a straight-line relationship.*

When a weight W, is applied to the lower end of a vertical steel wire and the resulting extension (e) of the wire is measured we know that Hooke's law is obeyed if the load is not too big. So $e = kW$ where k is a constant that is characteristic of the steel wire used. If different loads are used and a graph is plotted of extension e versus load W a straight line is expected that passes through ($W = 0$, $e = 0$). So the proportionality of e and W should be confirmed and it remains only for the value of k to be discovered.

Assuming that the graph line is as expected, the k is determined by measuring the slope of the e versus W graph (or the reciprocal of the slope if W were plotted as ordinate and e as abscissa).

If the unstretched length of the wire is L and its cross-section area is A then the Young modulus (Y or E) for the steel can be discovered from the formula $Y = \dfrac{WL}{Ae}$ or $Y = \dfrac{L}{Am}$ where m is the slope of the e versus W graph.

Here is another example.

The periodic time T (time for each complete swing) of a simple pendulum is given by the formula $T = 2\pi \sqrt{\dfrac{L}{g}}$ where L is the pendulum length, π is *pi* and g is the acceleration due to gravity. Measurements can be made of the period T for various values of length L.

We want to plot a graph to get a straight line (a *linear* relationship) that will confirm the formula for T and either confirm that the 2π is correct for the relationship, or more likely confirm or discover the value of g.

A single pair of L and T values could be recorded and used to calculate g from $\sqrt{g} = \dfrac{2\pi}{T}\sqrt{L}$

or $g = 4\pi^2 L/T^2$ but it is better to get several pairs of results to achieve an averaging of the results and consequently a more reliable answer for g. So we want a graph that will do this.

For the period we have $T = 2\pi\sqrt{\dfrac{L}{g}}$ or $T^2 = 4\pi^2\dfrac{L}{g}$ and clearly a graph of T^2 as ordinate versus L as abscissa should give a straight line and its gradient (m) should be $4\pi^2/g$ (*see Fig. 14b*). Hence $g = 4\pi^2/m$. The gradient m is the idealised value or averaged value of T^2/L. So g is obtained from a measurement of the gradient.

KEY FACT   *Formulae are often rearranged to produce a linear relationship between variables to be plotted.*

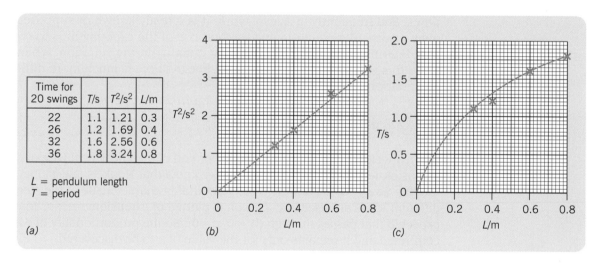

Time for 20 swings	T/s	T^2/s^2	L/m
22	1.1	1.21	0.3
26	1.2	1.69	0.4
32	1.6	2.56	0.6
36	1.8	3.24	0.8

L = pendulum length
T = period

(a)   (b)   (c)

Fig. 14 *Graphs for simple pendulum experiment.*

Further examples are given below.

Example

For a slow flow of liquid through a horizontal tube the volume per second (Q) passing through is given by $Q = \pi Pa^4/8L\eta$. When $P = hdg$ the equation becomes $Q = \pi hdga^4/8L\eta$.

The aim is to evaluate η. Different values of h can be used and the corresponding values of Q will be measured, while all the other quantities involved are constant.

(a)   What would be a suitable graph for the results?

(b)   If instead h stays unchanged and different Q values are obtained using various values for a, what would be a suitable graph to plot in this case?

Answer

(a) There is no need to rearrange the formula for Q. Q can be plotted (ordinate axis) versus h (abscissa) and a straight line should be obtained that passes through (0, 0) because $Q = \text{constant} \times h$. The constant is the gradient m of the graph and $m = \dfrac{\pi dg a^4}{8L\eta}$.

Rearranging this equation gives $\eta = \dfrac{\pi dg a^4}{8Lm}$. The gradient m is deduced from the graph, d, L and a are measured, g and π are known and η can be calculated.

(b) The formula for Q has the form $Q = m \times a^4$. With Q plotted as ordinate versus a^4 as abscissa the gradient is m and equals $\dfrac{\pi hdg}{8L\eta}$. So $\eta = \dfrac{\pi hdg}{8Lm}$ and can be calculated from the measurements.

Example

This example has the advantage of showing that measuring the intercept of a graph can be useful.

The formula $I = \dfrac{V}{\sqrt{R^2 + (2\pi fL)^2}}$ applies to the *alternating current I* that flows in a circuit containing inductance (L) and resistance (R) in series. V is the alternating (sinusoidal) voltage causing the current and f is its frequency. In a typical experiment f is varied and corresponding values of I are recorded. What graph should be plotted for a straight line to be obtained and how could measurements from the graph allow L and R to be determined?

Answer

The equation relating I and f needs to be in the form $y = mx$ or $y = mx + c$. Rearranging the formula for I gives $(2\pi fL)^2 + R^2 = \dfrac{V^2}{I^2}$ and further *transposition* gives $\dfrac{1}{I^2} = \dfrac{4\pi^2 L^2 f^2}{V^2} + \dfrac{R^2}{V^2}$.

This has the form $\dfrac{1}{I^2} = mf^2 + c$ and shows that plotting the reciprocal of I^2 versus f^2 will give a straight line.

The *gradient* will be m which is $\dfrac{4\pi^2 L^2}{V^2}$ so that $L = \sqrt{\dfrac{mV^2}{4\pi^2}}$ and L can be calculated from $\dfrac{V}{2\pi}\sqrt{m}$.

The *intercept* (on the y-axis) is c which equals R^2/V^2 and R can be calculated from $R = \sqrt{cV^2}$ or $V\sqrt{c}$.

Test Yourself

Exercise 9.4.1

For each of the following formulae suggest what quantities should be plotted and deduce a formula for calculating the required quantity or quantities from the gradient or intercept of the graph.

1 Force = mass × acceleration ($F = ma$). F is kept constant and is to be determined.

2 The frequency of the sound from a plucked string is given by $f = \dfrac{1}{2L}\sqrt{\dfrac{T}{m}}$ in which T is the tension in the string, L the string length and m is the mass per unit length. T and m are to be constant and L will be varied to give various frequencies. T is to be determined, m is known.

3 The energy radiated per second (i.e. the power radiated) from a hot object at absolute temperature T is given by $P = \sigma AT^4$. The σ and A are constants, A is known and σ is to be determined.

Area under a graph

We have seen that the gradient of a graph or an intercept will often tell us the value of a constant involved in the relationship described by the graph. The area beneath the graph line can sometimes be useful too. *Fig. 15(a)* shows a graph in which a constant velocity V exists for all values of the time t.

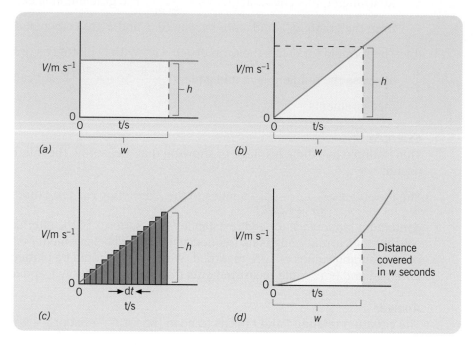

Fig. 15 *Area under a graph*

In this graph the *area under the graph* line between time $t = 0$ and $t = w$ is shaded. Its area is $h \times w$. Now if this area is calculated using not millimetres or 'squares of graph paper' for measuring h and w but using the same units as for the y and x-axes respectively then the area equals the product of velocity $\times$ time of travel and this equals the distance covered.

KEY FACT *The area under a velocity versus time graph equals the distance covered. For the area under a graph the unit is the product of the units of the x and y-axes.*

In *Fig. 15(b)* the velocity increases steadily with time, i.e. there is a **uniform** (steady or **constant**) acceleration. Does the area under the graph again tell us the distance covered? The answer is yes. As shown in *Fig. 15(c)* the area can be imagined as divided into a large number of thin vertical rectangles each having a small width δt (read as *delta t* and meaning a small increase in t). The area of any one of these rectangles is its velocity $\times \delta t$ and represents the distance covered in that short time. The total area under the graph is the total of all these rectangles and so is the total distance covered. A similar argument tells us that even when a graph line is curved as in *Fig. 15(d)* (acceleration increasing with time) the area under the graph still gives the distance travelled.

In *Fig. 15(b)* the triangular area is half of the rectangle having area $h \times w$ or final velocity $\times t$ which means that the triangle area is $\frac{final\ velocity}{2} \times t$ or average velocity $\times$ time. (Averages are discussed in Chapter 12.)

KEY FACT *For a triangle the area is half the width $\times$ the height.*

This rule applies to all triangles not just for a right-angled triangle as will be explained in Chapter 10.

For a small positive electric charge approaching a small positive charged object a graph can be plotted of the electric intensity E (the force that will act on each coulomb of the approaching charge) versus the distance d between the charges (see *Fig. 16*).

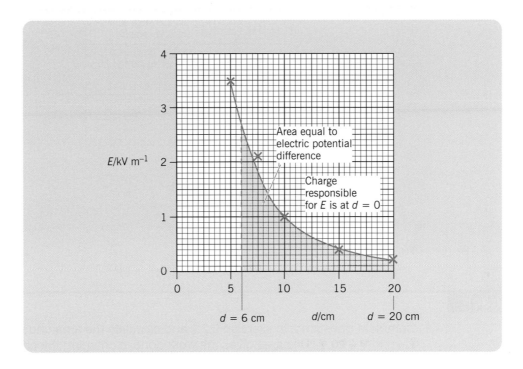

Fig. 16 *Electric potential and area beneath intensity versus distance graph.*

The area under the graph is then the work done on each coulomb of the approaching charge which is its electrical potential energy increase. If the charge starts from infinity (very large d) the electric potential energy gained per coulomb is the electric potential caused by the charged object at the place finally reached. Thus the area beneath the graph allows the potential for $d = 5$ cm, to be deduced.

The unit for the area beneath a graph (= product of x and y units) is an indication of the kind of quantity represented by the area under a graph. For example when force stretching a wire is plotted versus extension, the area unit is N m or J (or a multiple of it) and this is the unit for work or energy. Hence area represents energy stored.

For a graph drawn on squared paper an area can be measured approximately by first counting the number of squares in the area, e.g. using 1 cm by 1 cm squares. This area must then be converted to appropriate units according to the units of the x and y-axes. These are kV m^{-1} and cm (see question 3 in the next exercise and its hint).

In the above discussion w and h were used to describe the area under the graph of *Fig. 15(b)*. Instead we might have used v and t. In this case where v and t are now the *final* velocity and *final* time for the distance covered it is best to avoid v and t for labelling the graph axes and use the words *velocity* and *time* for the axes. This practice is used in *Fig. 17* where the pressure of a gas is plotted against the volume of the gas. The area under the graph gives the work done by the expanding gas as it pushes back a piston or the walls of its enclosure.

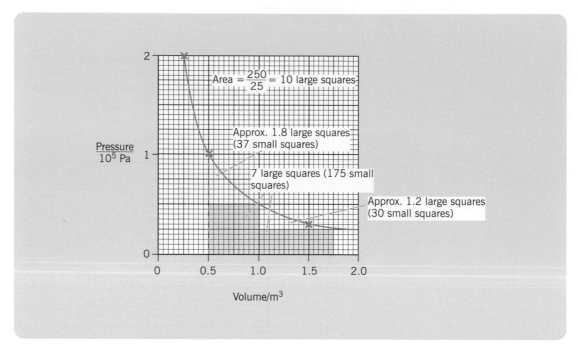

Fig. 17 *Area beneath a P versus V curve.*

1 Complete question 2 in exercise 9.2.1 and calculate the area under the graph from $V = 0$ to $V = 20$ V. This area gives the work done in charging the capacitor which is the energy finally stored in it.

HINT ▷ *Calculate work in joules by using SI units, so not mC but coulombs.*

2 If the power dissipated in a resistance (i.e. the electric energy converted to heat per second) is made to steadily increase with time what quantity does the area under the power versus time graph represent? $\left(\dfrac{Power = energy\ converted}{time}\right)$

3 Estimate the area under the curve in *Fig. 16* by counting squares and so estimate the electric potential difference between 6 cm and 20 cm from the charge causing this potential difference.

HINT ▷ *Between 8 and 10 cm² with each cm² of paper representing 0.5 kV m⁻¹ × 2.5 cm or* $0.5\ kV\ m^{-1} \times \dfrac{2.5}{100}\ m\ or\ 0.0125\ kV.$

9.5 Some special graphs

Graph of eˣ

Fig. 18 shows three graphs. One of these is the curve for $y = 2^x$. On this curve the slope at any chosen point is less than the y value at that point.

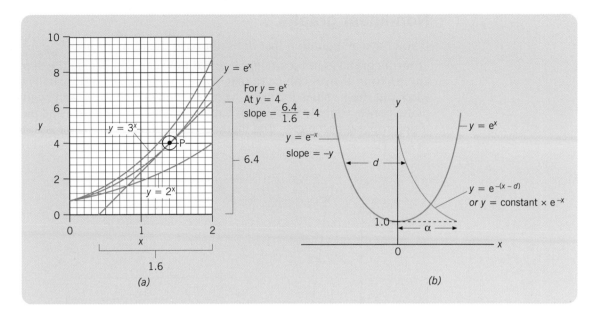

Fig. 18 Graphs including $y = e^x$

The slope at any point on the graph of $y = 3^x$ is *greater* than the y value of that point.

Between these two curves is the third curve which for any point such as the one marked P has a slope exactly equal to the y value there. The equation for this curve involves a number between 2 and 3, namely the **exponential function** denoted always by **e**. This e was first mentioned in Chapter 4. The equation for this special curve is $y = e^x$.

KEY FACT   *For $y = e^x$, gradient = y.*

If $y = e^{-x}$ is plotted you find that the gradient equals $-y$ and slopes downwards. The graph is shown in *Fig. 18(b)* which also shows $y = e^{-(x-d)}$ or $y = e^d e^{-x}$ or $y = $ constant $\times e^{-x}$.

It has already been mentioned that the value of e is 2.718 to 4 s.f. As a student there is no need for you to remember the 2.718. Obtaining and using the value of e with a calculator was explained in Chapter 4.

The equation $A = A_0 e^{-\lambda t}$ was met in Chapter 4. It describes radioactive decay. The A denotes **activity** which is the number of decays occurring per second. Related equations are $n = n_0 e^{-\lambda t}$ and $A = -\lambda n$, n being the number of atoms of radioactive substance.

By comparing $n = n_0 e^{-\lambda t}$ with $y = $ constant $\times e^{-x}$ we see that n falls as shown in *Fig. 18(b)* and, since the rate of change of y with x $\left(= \dfrac{dy}{dx} \right)$ equals $-y$, we have $\dfrac{dn}{d(\lambda t)} = -n$ or

$\dfrac{dn}{dt} = -\lambda n$. But $A = \dfrac{dn}{dt}$ so $A = -\lambda n$ as stated above.

Also, writing $-A/\lambda$ in place of n changes the formula for n into $A = A_0 e^{-\lambda t}$.

Non-linear graphs

If you have an equation that does not have the form $y = mx + c$, $y = \frac{2}{x^2}$ for example, then you know its graph is not a straight line but how can you decide the approximate shape of the curve? Your answer can be to take some simple values of x, usually 0, 1, 2, 3, 4, etc, and work out the y values for these. You can then sketch the graph by just drawing axes on plane paper, roughly marking in equal intervals on each axis and plotting your points as if on proper graph paper. The method can be very quick. The result for $y = 2/x^2$ is shown in *Fig. 19*.

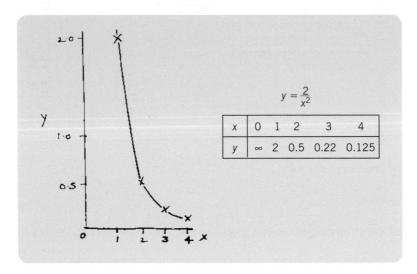

$$y = \frac{2}{x^2}$$

x	0	1	2	3	4
y	∞	2	0.5	0.22	0.125

Fig. 19 *Sketching a graph.*

Graphs concerning roots

A graph for a relationship $y = k\sqrt{x}$ (meaning $k \times \sqrt{x}$) or $y = kx^{0.5}$ will be a curve whereas $y = kx$ gives a straight line. For the curve, for the same k, the changes in y occurring after $x = 1$ are much less rapid than for $y = kx$. These features are shown for $k = 1$ in *Fig. 20*.

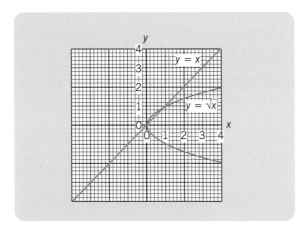

Fig. 20 *Graphs involving roots are compared.*

In *Fig. 20* it may also be seen that $y = \sqrt{x}$ gives no values for negative x values but two answers for positive x whereas $y = \sqrt[3]{x}$ gives only + answers for positive x and negative values for negative x. These results have already been discussed on page 80 under the heading

Positive and negative roots in Chapter 6.

For a $y = \sqrt{x}$ graph where x is, measured in metres, the unit for y would be metre$^{0.5}$. What would the unit be for $y = x^4$ if the unit for x were cm? (The answer is cm^4.)

Exam Questions

Exam type questions to test understanding of Chapter 9

Exercise 9.4.3

1 The graph seen in *Fig. 21* shows the relation between the product *pressure × volume*, PV, and temperature θ, in degrees Celsius for 1 mol of an ideal gas for which the molar gas constant is R.

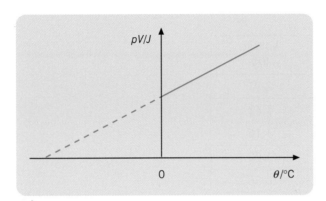

Fig. 21

Which one of the following expressions gives the gradient of this graph?

A $\frac{1}{273}$ **B** $\frac{PV}{\theta}$ **C** $\frac{PV}{(\theta - 273)}$ **D** R

(AQA 1999)

> **HINT**   *PV = RT is an important formula in physics.*

2 The graph in *Fig. 22* shows the charge stored in a capacitor as the voltage across it is varied.

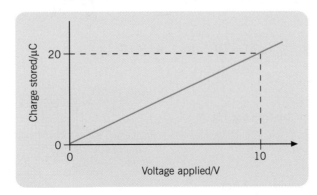

Fig. 22

The energy stored, in μJ, when the potential difference is 5 V, is

A 25 **B** 50 **C** 100 **D** 200

(AQA 2000)

HINT>

Energy is area beneath graph. For 10 V area is half of rectangle area.

3 It is suggested that the turn-on time, T_{on}, for a liquid crystal display is given by the equation

$$T_{on} = \frac{k\eta d^2}{V^2}$$

where η is the viscosity, V the voltage applied, d the thickness of crystal, and k is a constant.

Data showing how the turn-on time T_{on} depends on the voltage V is provided in the table

Table 1

Turn-on time T_{on}/ms	Voltage V/V
5	2.01
10	1.42
15	1.16
20	1.00
27	0.86

(a) On the graph paper provided, plot a suitable graph to test the relationship suggested between T_{on} and V. Record the results of any calculations that you perform by adding to a copy of the table.

(b) Discuss whether or not your graph confirms the suggested relationship between T_{on} and V.

(c) Use your graph to calculate a value of the constant k. The viscosity of the liquid crystal substance is 0.072 Pa s and its thickness is 6.0 μm.

(d) A second experiment is to be carried out to test the relationship between the turn-on time T_{on} and the thickness d. Suggest a possible pair of axes to obtain a straight line graph from measurements of T_{on} and d.

List all the possible variables that would need to be kept constant.

(Edexcel 01)

HINT>

Pa s is S.I. unit for η. 1 μm = 10^{-6} m, 1 ms = 10^{-3} s.

Answers to Test Yourself Questions

Exercise 9.2.1
1 (a) 0.15   (b) 0.05 A V^{-1}
2 0.05 mC V^{-1} (or 50 microfarad). See Fig. A1.

Exercise 9.2.2
1 $P = 1.2\theta + 75$
2 $u = 30$ m s^{-1} approximately, $a = 2.5$ m s^{-2} approximately. See Fig. A2.

Exercise 9.3.1
1 20. See Fig. A3 for graph.
2 20 m s^{-1}, 40 m s^{-1}. See Fig. A4 for graph.

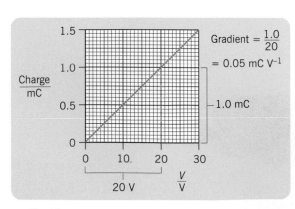

Fig. A1

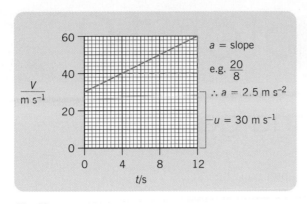

Fig. A2

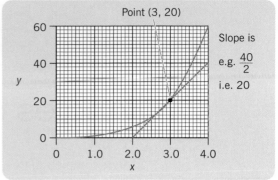

Fig. A3

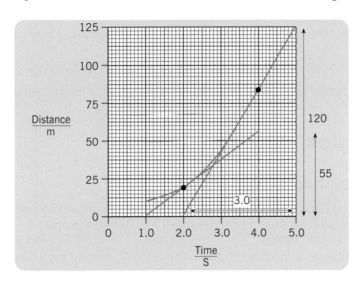

Fig. A4

Exercise 9.4.1

1 Plot a versus $1/m$, F = gradient or $1/a$ versus m, $F = 1/\text{gradient}$.

2 Plot f^2 versus $1/L^2$, gradient $= f^2L^2$, $T = 4m \times$ gradient. For $1/f$ versus L, $T = \frac{4m}{\text{gradient}^2}$. For f versus $1/L$, $T = 4m \times \text{gradient}^2$.

3 Plot P versus T^4 and gradient $= \sigma A$, $\sigma = \frac{\text{gradient}}{A}$.

Exercise 9.4.2

1 1.0×10^{-2} J or 10 mJ
2 Energy converted.
3 Between 0.10 and 0.13 kV.

Chapter 10

Angles and geometry

After completing this chapter you should:

- *be familiar with the naming of angles*
- *know the properties of triangles and circles*
- *know how to use the Pythagoras formula*
- *understand the meaning of the radian and its importance for movement on a circular path*
- *be able to calculate volumes of rectangular blocks, cylinders and spheres.*

(Sines and cosines of angles and other trigonometry topics are the subject of Chapter 11.)

10.1 Angles

Angles measured in degrees

A shaft may be turned by an electric motor or a satellite may be moving around the Earth, you may want to calculate the angle the object has moved through.

In *Fig. 1(a)* a rotation of the line OA about the point O starting from where it is coincident with OB will make it coincident with OB again after a complete **revolution**. This rotation of 1 revolution is called 360 **degrees** (written as 360°).

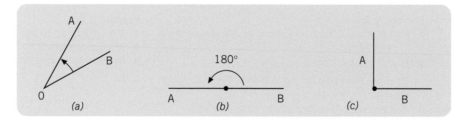

Fig. 1 *Angles.*

The separation of the lines by the rotation is described as an angle. So one revolution is an angle of 360°.

KEY FACT   *1 revolution is 360°.*

In *Fig. 1(b)* OA has only turned through half a revolution or 180° and in *Fig. 1(c)* only through a quarter of a revolution or 90°. A 90° angle is called a **right-angle** and the symbol used for an angle is ∠.

KEY FACT   *∠90° is a right-angle.*

When two lines form a right-angle they are **perpendicular** or **normal** to each other.

Fig. 2(a) shows how an angle in a diagram is marked with a curved line and how the angle size is shown.

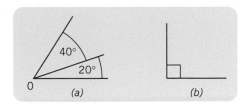

Fig. 2(b) shows how a right-angle is marked with a line that forms a small square in the angle.

Fig. 2 *Marking angles in diagrams.*

In diagrams which show a triangle (see *Fig. 3*) it is usual to use capital letters (A rather than a) to name places such as the angles of the triangle. A, B and C are an obvious choice for a triangle. When letters are used to describe lengths or distances it is customary to choose lower case letters (a not A).

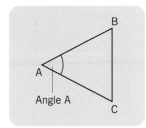

When an angle is not named in this way lower case letters are often used and *Greek letters* are popular, especially θ (*theta*), ϕ (*phi*), α (*alpha*), β (*beta*) and γ (*gamma*).

Fig. 3 *Naming angles.*

KEY FACT   *Capital letters for places. Lower case for lengths.*

In *Fig. 4(a)*, *(b)*, and *(c)* respectively are examples of an **acute angle** (less than 90°), an **obtuse angle** (between 90° and 180°) and a **reflex angle** (greater than 180°).

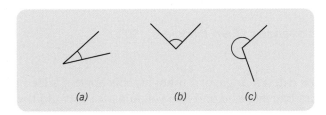

Fig. 4 *Names of some angles. (a) An acute angle. (b) An obtuse angle. (c) A reflex angle.*

KEY FACT   *Remember the terms* acute, obtuse *and* reflex.

Angles of a triangle

You will need to know a few facts about triangles not only when you are making calculations but also when you are learning about combinations of forces, alternating currents and interference of waves.

A triangle has three angles inside it, three **interior** angles. *Fig. 5* shows a triangle with a set of **exterior angles** marked in as well.

In *Fig. 5(a)* imagine a little person walking from A clockwise around the edge of the triangle back to A. Three right turns are made through the three exterior angles and the effect is a complete rotation through 360°. So the sum of a set of exterior angles is 360°.

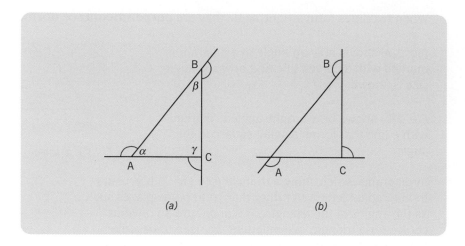

Fig. 5 *Exterior angles of a triangle (a) Exterior angles (b) The other set of exterior angles*

KEY FACT   *The sum of the exterior angles of a triangle is 360°.*

The interior angles of the *triangle* in *Fig. 5(a)* are marked as α, β and γ. Now α plus the exterior angle at A add to 180° and so the three exterior angles plus the three interior angles of the triangle add to $3 \times 180 = 540°$. Subtracting the 360° for the exterior angles gives the sum of the interior angles as 180°.

KEY FACT   *The sum of the interior angles of any triangle is 180°.*

Fig. 6 shows just one exterior angle of a triangle. This angle if added to angle γ would give a sum of 180°. But adding angles α and β to γ would give 180°. So the exterior angle equals $\alpha + \beta$.

KEY FACT   *An exterior angle of a triangle equals the sum of the opposite interior angles.*

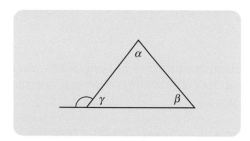

Fig. 6 *Exterior angle equal to the sum of the opposite interior angles.*

Example
Fig. 7 shows a vehicle at rest on a very slippery slope, held by a rope. The slope makes an angle of 5° with the horizontal. What is the angle between the reaction force of the slope and the vertical?

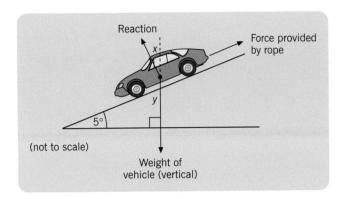

Fig. 7

Answer

On a very slippery slope the reaction force is perpendicular to the surface as shown. The triangle containing the 5° has for its other angles the angle y and a right-angle. This means that the total of the interior angles is $5 + y + 90$ and equals 180 and so $y = 85°$.

To find the required angle, x, we have $x + 90 + y$ making an angle of 180° and x works out to 5°.

Test Yourself	Exercise 10.1.1

1 In *Fig. 8* angle x plus angle $y = 70°$. What are the values of angles x, y and z?

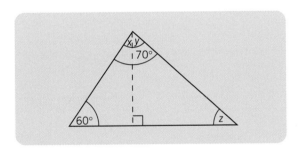

Fig. 8

2 In *Fig. 9* what is the value of angle x?

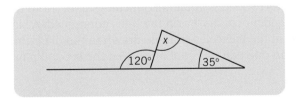

Fig. 9

3 *Fig. 10* shows a hinged trap-door which is held in a horizontal position by hinges attaching it to a vertical wall and held by a rope suitably tied to its left-hand edge and to a point on the wall as shown.

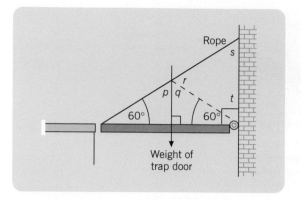

Fig. 10

The weight of the trap-door acts as if it were concentrated at the middle of the trapdoor. Two angles are shown as 60°. Work out the angles *p*, *q*, *r*, *s* and *t*.

HINTS AND TIPS For *p* consider right-angled triangle. For *s* consider large right-angled triangle.

10.2 Isosceles and equilateral, congruent and similar triangles

Isosceles and equilateral triangles are illustrated in *Fig. 11*.

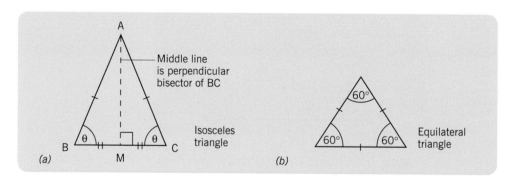

Fig. 11 *Some special triangles (a) An isosceles triangle (b) An equilateral triangle*

An **isosceles** triangle has two sides equal. (*Iso* means *same*). Note that it is common to show equal lengths by placing identical marks on the lines concerned. A single slash is shown on each of the equal sides in the isosceles triangle drawn in the diagram. If any diagram (see *Fig. 11(a)*) has more than one pair of equal lines then the second pair can be marked with two slashes each.

The special property of an isosceles triangle is that it is symmetrical about a middle line perpendicular to the odd side and passing through the opposite corner (**vertex**) as shown. This statement means that the diagram on one side of the line is a mirror image of the other side. A consequence is that the angles on each side are equal.

KEY FACT *An isosceles triangle has two equal angles.*

While considering the isosceles triangle it is worth mentioning that the middle line in *Fig. 11* divides the side BC into two equal parts, two halves, and is therefore a **bisector** of that side. In this particular case the line is also perpendicular to BC and so is the **perpendicular bisector** of BC.

An **equilateral** triangle has all its three sides equal (*lateral* referring to *sides*). Since the three corners of the triangle are identical the angles are all equal and are each 60° so that the three total to 180° for the interior angles of the triangle. Of course an equilateral triangle could equally well be called an **equiangular triangle**.

> **KEY FACT**   *An equilateral triangle has each angle equal to 60°.*

When two or more triangles are compared they may be identical in shape and size as shown in *Fig. 12(a)*. They are then said to be **congruent**. They would fit upon each other.

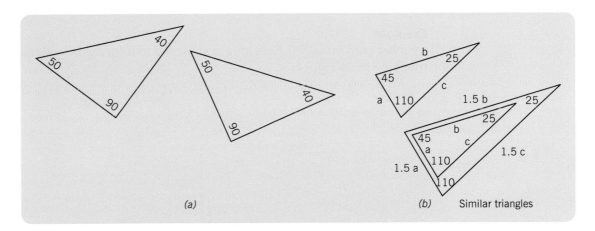

Fig. 12 *(a) Congruent triangles (b) Similar triangles*

In *Fig. 12(b)* the three triangles are not the same size but they do have the same shape because they contain the same angles and they are described as **similar** triangles. These triangles have the property that corresponding sides are in the same ratio, the ratio being 1.5 to 1 in *Fig. 11(b)*.

> **KEY FACT**   *Congruent triangles have the same shape and size. Similar triangles have the same shape (same corresponding angles) and corresponding sides in the same ratio.*

10.3 Pythagoras' formula

For any triangle that contains a right-angle the lengths a, b and c of the triangle's sides fit the equation $a^2 = b^2 + c^2$. This is the *Pythagoras* formula.

> **KEY FACT**   *Pythagoras' formula $a^2 = b^2 + c^2$.*

In this equation the side of length a must be the longest side and this, of course, is the side opposite the right-angle (see *Fig. 13*). This side is given the name of **hypotenuse**.

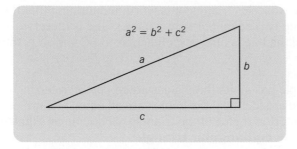

Fig. 13 *The Pythagoras formula.*

A proof of this equation is given in Appendix II.

The Pythagoras formula was used in Chapter 3 on page 42 when *Using a product of brackets in physics* was explained. This formula is particularly useful for vector calculations and for deriving formulae concerning alternating currents.

Example

A rubber cord initially 0.12 m long is stretched to form the longest side AC of the right-angled triangle ABC in *Fig. 14*.

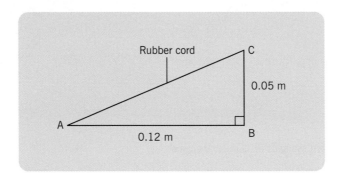

Fig. 14

If each centimetre of extension requires a force of 0.20 newton what is the tension in the stretched cord?

Answer

Applying the Pythagoras rule to the triangle,

$$AC^2 = 0.12^2 + 0.05^2 = 0.0144 + 0.0025$$
$$= 0.0169$$
$$\therefore \quad AC = \sqrt{0.0169} = 0.13 \text{ m}$$

The extension is $0.13 - 0.12 = 0.01$ m or approximately 10 cm. (Note that two significant figure accuracy is not really justified here because of the subtraction of similar values.)

The tension force (i.e. force associated with stretching) is approximately

$$10 \text{ cm} \times 0.20 \text{ N cm}^{-1} \text{ or } 2.0 \text{ N.}$$

Test Yourself

1 The longest side of a certain right-angled triangle is 13 cm and the shortest side measures 5.0 cm. Calculate the length of the third side.

2 An electric charge of size 1.0×10^{-9} coulomb is placed at point A of a square ABCD of side 0.20 metre. Calculate the electric potential at C.

(Potential V in volts $= \dfrac{Q}{4\pi\varepsilon R}$ where R is the distance in metres between the point

concerned and the position of the charge Q coulomb. The value for $\dfrac{1}{4\pi\varepsilon}$ in SI units can

be taken as 9.0×10^{9}.)

HINT — *Find AC = 0.2828 m.*

Area of a triangle

In Chapter 3 the *area* of a triangle was calculated from half its base times its height because the triangle was a right-angled triangle and so was clearly half of a rectangle with the same height and width. *Fig. 15(a)* reminds us of this.

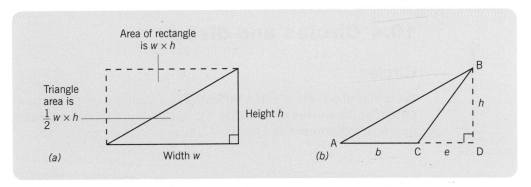

Fig. 15 *Areas of triangles: (a) A right-angled triangle (b) Any triangle.*

We now consider a triangle such as that shown in *Fig. 15(b)* which has no right-angle. The area of triangle ABC = area of triangle ABD minus area of triangle BDC

$$= \frac{1}{2}(b + e)h - \frac{1}{2}eh = \frac{1}{2}bh$$

This means that the same formula applies although **base** is usually preferred to *width* now that the triangle has no right-angle. The base is the side perpendicular to the height and from which the height is measured.

KEY FACT — *The area of any triangle $= \frac{1}{2} \times base \times height$.*

Test Yourself

1 Calculate the area of a right-angled triangle with shortest sides measuring 3 cm and 6 cm.

2 (a) Calculate the area (A) beneath the graph of $v = 2t$ from $t = 0$ to $t = 3$.
　(b) State the unit for A in part (a) given that the units for v and t are m s^{-1} and s respectively.

Quadrilaterals

A **quadrilateral** is a shape with four sides. *Fig. 16* shows some examples. **Parallelograms** have both pairs of opposite sides parallel. For a **rectangle** all four angles are right-angles, while a **square** of course has equal sides and four right-angles (and of course opposite sides parallel).

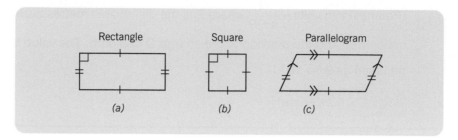

Fig. 16 *Quadrilaterals.*

The area of a rectangle is worked out by multiplying its length and width and this has been explained already.

10.4 Circles and discs

Circles

Every part of a circle is at the same distance from the centre of the circle, this distance being known as the **radius** of the circle. The distance across the circle through the centre is called the circle's **diameter** and this is of course twice the radius.

KEY FACT *Diameter of a circle = 2 × radius.*

A radius and a diameter of a circle are shown in *Fig. 17* and a chord and a sector are also included.

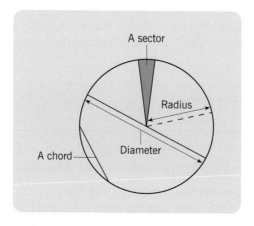

Fig. 17 *Names associated with circles.*

The length of the curved line that forms a circle is the circle's **circumference**. This distance is always a little more than three times the diameter of the circle or disc, in fact 3.142 times, but you don't need to memorise this figure. Just remember that it is called *pi* and is always denoted by the *Greek symbol* π. So the circumference = $\pi \times d = 2\pi r$ if we use d for diameter and r for radius.

KEY FACT   *Circumference = $2\pi r$.*

Measure the circumference and diameter of a plate with a tape measure and confirm that it approximately equals $2\pi r$ or $\pi \times$ diameter.

A portion of a circle's circumference is called an **arc**.

On your calculator you should find a key for π. It is the second function of the EXP key. Pressing the SHIFT key first, then the key for π and = will give 3.141592654.

Consider a circle having a radius of 7 cm. Use the keys

$$\boxed{2}\ \boxed{\times}\ \boxed{\text{SHIFT}}\ \boxed{\pi}\ \boxed{\times}\ \boxed{7}\ \boxed{=}$$

and you should get 4.398×10^1 cm, close to 44 cm. To make a rough check use 3 for π and we have $2 \times 3 \times 7$ which equals 42.

Physics calculations often deal with things moving along circular paths. Examples are a tooth (or cog) on the circumference of a wheel; an object being whirled round at the end of a string; a vehicle moving round a circular track and an electron in orbit around its nucleus. Later in this chapter we will deal with the speeds of rotation.

A disc

A flat area whose edge is a circle is a disc. The area of a disc is equal to $\pi \times$ radius^2. This formula is important when you calculate the resistivity of a metal wire or the flux density in a coil. It is often described as the formula for the **area of a circle**, meaning the area *within* a circle.

If you cut across a round wire the cut surface, the **cross-section** is a disc. The area of the cross-section is important in electrical calculations.

The formula for the area of a disc can be derived as follows. If the sector of the circle in *Fig. 17* only used a small bit of the circumference it would be a thin triangle the arc being like a straight line. Its height is the radius of the circle and its base is its bit of the circumference. We next picture the whole disc as made up of a large number of such thin triangles, each having area equal to half base times height. The *area of the disc* equals the total of the sector areas and so is half times height times whole circumference and this amounts to $\dfrac{1}{2} \times r \times 2\pi r = \pi r^2$.

KEY FACT   *The area of a disc (or a circle) = πr^2.*

Example

The number of free electrons per metre cubed in copper is approximately 10^{29}. Calculate the number of such electrons per metre length of a copper wire having a diameter of 1.0 mm (Electron charge $= 1.6 \times 10^{-19}$ C.)

Answer

Volume = length × area of cross-section
Area of cross-section $= \pi r^2 = \pi \times (0.5 \times 10^{-3})^2 = \pi \times 2.5 \times 10^{-7}$ m^2
$\therefore$ Volume $= 1 \times \pi \times 2.5 \times 10^{-7}$ m^3
Electrons per metre length $= 10^{29} \times 2.5\pi \times 10^{-7} = 2.5\pi \times 10^{22}$
$= 7.855 \times 10^{22} = 7.9 \times 10^{22}$ m^{-1}

Test Yourself

Exercise 10.4.1

Take π to be 3.142.

1 Calculate the area of a circle that has a radius of 0.30 m.

2 A flat circular coil of wire lies perpendicular to a uniform magnetic flux density (*B*) of 0.02 tesla. The coil radius is 1.5 cm. Calculate the magnetic flux through the coil.

HINTS AND TIPS

The formula required is $\varphi = BA$ where φ is not an angle but the magnetic flux in weber (Wb), A is the coil area in m^2. Convert cm to metre.

3 A water tank in the shape of a vertical cylinder having a diameter of 0.80 m contains water to a depth of 0.50 m. What volume of water does it contain?

HINT

Radius is 0.40 m.

10.5 Three-dimensional shapes

Why *three*-dimensional?

Rectangles or triangles are examples of flat shapes (or flat **figures**) – each has all its parts in the same plane. They are two-dimensional figures because any point on the figure requires two measurements to specify its position. Here the term **dimensions** is used a little differently from its previous use in Chapter 5. Here it means measurements in different directions rather than measurements of different types.

Now a box is a **3-dimensional** object and in *Fig. 18(a)* the chosen three dimensions are the *x*, *y* and *z*-axes and these axes have been placed to coincide with three edges of a box.

A solid block having the same shape as that in *Fig. 18(a)* can be called a *rectangular block* and **cuboid** is the mathematical term for either the box or rectangular block. If the cuboid has all edges of equal length then of course it is a cube. The volume of a cuboid is given by length × width × depth as explained in Chapter 5.

If a circle is made to spin rapidly about a diameter it is seen as a sphere (*see Fig. 18(b)*). A **cylinder** has a cross-section which is a disc or circle (*see Fig. 18(c)*).

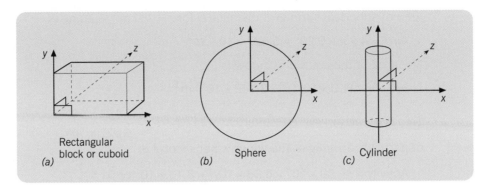

Fig. 18 *3-dimensional shapes.*

Spheres

The Earth and the planets and everyday objects such as light bulbs approximate to *spheres* and are met in physics calculations.

Two important facts need to be remembered about spheres. These are

KEY FACT *Volume of a sphere $= \frac{4}{3}\pi r^3$ and surface area of a sphere $= 4\pi r^2$ (r being the sphere's radius).*

Example
Calculate the mass of the planet Mars given the following data.

Density $= 3.9 \text{ g cm}^{-3}$ Radius $= 3.4 \times 10^6 \text{ m}$ (These are mean values.)

Answer

$$\text{Density} = \frac{\text{mass}}{\text{volume}}$$

$\therefore$ mass $=$ density $\times$ volume

$$\text{Volume} = \frac{4}{3}\pi r^3 = \frac{4}{3} \times \pi (3.4 \times 10^6)^3 = 1.646 \times 10^{20} \text{ m}^3$$

Density in g cm^{-3} $= 3.9$ and in kg m^{-3} $= 3.9 \times 10^{-3} \times 10^6 = 3.9 \times 10^3$ kg m^{-3}.

$\therefore$ mass $= 3.9 \times 10^3 \times 1.646 \times 10^{20}$ kg or 6.4×10^{23} kg.

Example
Calculate the number of beta particles arriving per second at the window of a geiger tube given that the radiation comes from a small radioactive source having a beta activity of 50 kBq at a distance of 20 cm. The window is placed perpendicular to the incident radiation and has a radius of 1.0 cm.

Answer
The number of beta particles emitted per second equals the activity in bequerels which is

50×10^3.

Assuming the radiation to be emitted equally in all radial directions it will, at a distance r, be spread over an imaginary spherical surface having an area $4\pi r^2$, and the counter window is to be a small part of this surface.

Area of spherical surface at distance r, for $r = 20$ cm $= 0.20$ m, is

$$4\pi r^2 = 4 \times \pi \times 0.20^2 = 5.027 \times 10^{-1} \text{ m}^2$$

The area of the counter window is

$$\pi r^2 = \pi \times (1.0 \times 10^{-2})^2 = 3.142 \times 10^{-4} \text{ m}^2$$

This is a fraction f of the spherical area where $f = \dfrac{3.142 \times 10^{-4}}{5.027 \times 10^{-1}} = 6.26 \times 10^{-4}$ and the number of particles arriving at the window per second is

$$\text{Activity} \times f = 50 \times 10^3 \times 6.26 \times 10^{-4} = 3.13 \times 10^1 \text{ or } 31 \text{ s}^{-1}$$

Test Yourself

Exercise 10.5.1

Take π as 3.142 or use the key for π on your calculator.

1 Calculate the volume of a tungsten-filament light bulb that is spherical with a radius of 3.0 cm.

2 (a) Calculate the power radiated by the sun using the data given below. Absolute surface temperature (T) = 6000 K, radius (r) = 7.0×10^5 km, Stefan constant (σ) = 5.8×10^{-8} W m^{-2} K^{-4}.

HINT *Formula needed is power $P = \sigma A T^4$ where A is the area of the radiating surface.*

(b) Calculate the power received per metre squared on the Earth's surface when the Sun is overhead given that the Earth is 15×10^7 km from the Sun.
(c) Calculate the total power received by the Earth from the Sun given that the Earth's radius is 6400 km.

HINTS AND TIPS *(a) Change km to m (b) Sun sees Earth as a disc. See worked example above.*

Cylinders

The volume of a *cylinder* is given by the formula $V = \pi r^2 h$ where r is the radius of the cylinder's base and h is the cylinder's height. This formula follows from the rule used earlier for the volume of a cuboid that a volume equals the base area times the height, provided that the sides are vertical.

KEY FACT *Volume of cylinder* $= \pi r^2 h$.

Example
Calculate the volume of a 0.50 m length of wire having a radius of 0.50 mm.

Answer
0.50 mm $= 0.50 \times 10^{-3}$ m   Let the wire length be L.

The volume $= \pi r^2 L = \pi \times (0.50 \times 10^{-3})^2 \times 0.50 = 4.0 \times 10^{-7}$ m^3.

---

1 Calculate the pressure beneath a vertical, cylindrical pillar having a radius of 50 cm, height 3.0 m and density 2500 kg m^{-3}. Take g as 10 m s^{-2} or 10 N kg^{-1}. $\pi = 3.142$

HINTS AND TIPS

Pressure $= \dfrac{weight}{area} = \dfrac{mass \times g}{area}$. *Work out volume, then mass, then weight, then pressure or use formula P = height × density × g. V = $\pi R^2 h$, m = V × density. Weight (the force) = mg.*

10.6 Angles measured in radians

The size of a radian

Radians are an alternative to degrees for angle measurement and are most useful when we are making calculations on rotating objects.

A *radian* is an angle such that just over six radians make up one revolution. The exact number is 2π and this is close to $2 \times 3.142 = 6.284$. The choice of 2π has been made to obtain many advantages over the degree.

KEY FACT   *1 revolution = 2π radian.*

The radian is the *SI unit* for angle measurement. It is close to $57.3°$. The abbreviation for radian is **rad**. The number of degrees per radian is $\frac{360}{2\pi}$ and the number of radians per degree is $\frac{2\pi}{360}$.

When otherwise working with degrees (in *degree mode*), if you want to use an angle in radians you can enter the radian value and then use a key labelled DRG. For example sine of 0.5 radian can be obtained by using sin 0.5 SHIFT DRG followed by choice of R for radian from the menu shown on screen. Then pressing = gives 0.479425538. When the screen is cleared (using the AC key) the calculator returns to working in degrees for subsequent calculations.

If any circle is drawn with its centre coinciding with the origin of a 1 radian angle then the length of arc of the circle that lies within the angle is equal to the circle's radius (see *Fig. 19*).

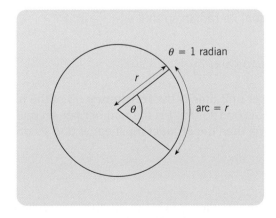

Fig. 19 *The arc subtended by 1 radian.*

KEY FACT   *The arc of a circle subtended by 1 radian equals the circle's radius.* **Subtended** *means 'opposite to'.*

The number of radians in an angle (θ) may be found by dividing the arc subtended by the angle by the radius (r) of the arc.

KEY FACT $\theta = arc/r$.

(This statement follows from the fact that 1 radian has an arc equal to r so that θ radian has an arc of $\theta \times r$ and hence $\theta = \frac{arc}{r}$. Otherwise you could say that θ is a fraction $\frac{\theta}{2\pi}$ of a revolution and so subtends an arc which is a fraction $\frac{\theta}{2\pi}$ of the circumference $2\pi r$ and so is $r\theta$.

Describing rotations

For any rotation the quickness with which it is occurring can be described by the angle moved through per second. This quantity is called the **angular velocity** or **angular speed**. It is denoted by the symbol ω (which is the lower case of Greek Ω) and is always measured in radian per second. ω is called *omega*.

KEY FACT   *Angular velocity (ω) =* $\dfrac{angle\ turned\ through}{time\ taken}$ *or* $\omega = \dfrac{\theta}{t}$.

A small object rotating or a point on a larger rotating object moves in a circular path. The distance covered per second along this path, i.e. the arc covered per second, is the **speed of rotation**.

KEY FACT   *Speed =* $\dfrac{arc}{time} = \dfrac{r\theta}{t} = r\omega$.

This *speed* can be constant, and will be in your calculations, although the velocity v is certainly changing as its direction changes for each point around the circular path. However, as regards the size of the speed and velocity they are the same and the abbreviation v is used for both.

While discussing angles of rotation it is appropriate to mention **period of rotation** denoted usually by T. This is the time for a complete revolution so $v = \frac{arc}{t}$ can be written as $v = \dfrac{2\pi r}{T}$. Also T is the time for θ to equal 2π radian so that $\omega = \dfrac{2\pi}{T}$ and $T = \dfrac{2\pi}{\omega}$.

KEY FACT   *Period (T) = time for 1 revolution and* $T = \frac{2\pi}{\omega}$.

The number of revolutions per unit time is the **frequency** (f) of the rotation. The preferred unit of time is of course the second so that f is the number per second. If f is multiplied by the time taken for each revolution then the product must equal one second, i.e. $f \times T = 1$ or $f = \frac{1}{T}$.

KEY FACT   *Frequency (f) =* $\frac{1}{T}$.

The obvious unit for measuring frequency is s^{-1} (a number *per second*). This is the SI unit for frequency but it is given a special name, **hertz** (abbreviation Hz).

The conditions that produce circular motions and the mathematics related to these are discussed in Chapter 15.

Test Yourself

Exercise 10.6.1

Use either your calculator for π or take it as 3.142.

1 Convert each of the following angles in degrees into radians:

   (a) 180°   (b) 60°   (c) 25°   (d) 2°

2 Convert from radians to degrees:

   (a) $\dfrac{\pi}{2}$ (b) $\dfrac{\pi}{5}$ (c) 0.30   (d) 8   (e) 5

3 A wheel rotates at a frequency of 20 Hz. What is its angular velocity?

Exam Questions

Exam type questions to test understanding of Chapter 10

Exercise 10.6.2
Take π to be 3.142.

1 The earth orbits the sun along an approximately circular path having a radius of 150×10^6 km and each orbit takes about 365 days to complete. Calculate (a) the speed of this movement and (b) the angular velocity.

2 The moon is approximately 3.8×10^5 km from the Earth and the diameter of the Moon is approximately 1.7×10^3 km. Calculate a value for the angle in radians subtended by the Moon from the Earth.

HINT ▷   *Angle = arc/radius = moon diameter/moon distance.*

3 A small 60 W filament lamp changes electrical energy to light with an efficiency of 12%. Calculate the light intensity produced by the lamp at a point 3.5 m from the filament.

(Edexcel 2001, part question)

HINT ▷   *Intensity* $= \dfrac{power\ in\ watts}{4\pi R^2}$ *where R is distance from source to place concerned.*
   Answer in watt per metre².

4 (a) State the number of protons and the number of neutrons in $^{14}_{6}C$.
   (b) The mass of one nucleus of $^{14}_{6}C = 2.34 \times 10^{-26}$ kg. The nucleus of carbon-14 has a radius of 2.70×10^{-15} m. Show that the volume of a carbon-14 nucleus is about 8×10^{-44} m³.
   (c) Determine the density of this nucleus.
   (d) How does your answer compare with the densities of everyday materials?
   (e) Carbon-14 is a radioisotope with a half-life of 5700 years. What is meant by the term half-life?
   (f) Calculate the decay constant of carbon-14 in s⁻¹.

(Edexcel 2001)

HINTS AND TIPS

6 protons, 14 is the number of protons plus neutrons. Volume $= \dfrac{4}{3}\pi R^3$.

Density = mass/volume. No maths in parts (d) and (e). Decay constant $\lambda = 0.69/T$ where

T is half-life. $(x \times 10^{-15})^3 = x^3 \times 10^{-45}$

Answers to Test Yourself Questions

Exercise 10.1.1
1 (a) 80   (b) alternate
2 (a) 360°   (b) 90°   (c) 90°
3 $p = 30°$, $q = 30°$, $r = 120°$, $s = 30°$, $t = 30°$

Exercise 10.3.1
1 12 cm
2 32 V
3 (a) 9.0×10^4 V or 90 kV   (b) 1.4×10^{-14} J
   (c) 1.8×10^8 m s^{-1}

Exercise 10.3.2
1 9 cm^2
2 (a) 9   (b) 9 m

Exercise 10.4.1
1 0.28 m^2

2 0.14×10^{-4} Wb
3 0.25 m^3

Exercise 10.5.1
1 1.1×10^2 cm^3 or 1.1×10^{-4} m^3
2 (a) 4.6×10^{-26} (b) 1.6 kW per m^2 (c) 2.1×10^{17} W
   (d) No loss of energy in source or atmosphere. Earth a perfect sphere.

Exercise 10.5.2
1 75 kPa

Exercise 10.6.1
1 (a) π radian or 3.14 radian
   (b) $\frac{\pi}{3}$ or 1.05 radian or 1.0 radian
   (c) 0.436 radian or 0.44 radian   (d) 0.035 radian
2 (a) 90°   (b) 36°   (c) 17°   (d) 458°   (e) 286°
3 126 rad s^{-1}

Chapter 11

Trigonometry

After completing this chapter you should

- *know the meanings of three trigonometric ratios, namely the sine, cosine and tangent ratios*
- *know relationships between these ratios that are needed for A level physics*
- *know what a vector is and be able to find the resultant of two vectors*
- *be able to make calculations on forces and directions for an object in equilibrium.*

11.1 Trigonometric ratios – what are they?

If a sledge is pulled by a rope the force moving the sledge forward is less than the pull by the rope because of the angle the rope makes with the forward direction. **Trigonometry** (the mathematics that uses trigonometric ratios) is needed to calculate the effective force.

Suppose that an angle, θ (see *Fig. 1*) is seen as an angle in a right-angled triangle. The length of the longest side of the triangle is called the **hypotenuse** and we denote this by a. The length of the side opposite the angle θ is denoted by b and the remaining side has a length c. The ratio b/a is the same regardless of how big the triangle is. So in the bigger triangle of *Fig. 1* the ratio (shown as $\dfrac{b'}{a'}$ in the diagram) has the same value as the ratio $\frac{b}{a}$

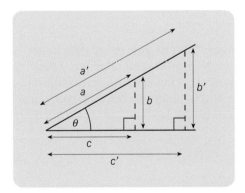

Fig. 1 *Trigonometric ratios.*

The name given to the ratio of opposite side divided by hypotenuse is **sine** θ (pronounced as 'sign theta') and the abbreviation for sine is **sin**. Note that sin is not pronounced as it looks but as 'sine'.

There are two more important **trigonometric ratios** that you need to be familiar with. These are **cosine** and **tangent**. Their abbreviations are **cos** and **tan**. The *co* of cosine is pronounced as in code but *cos* is pronounced like the cos in costume.

Cos θ is the ratio $\frac{c}{a}$ (see *Fig. 1*) of the side next to the angle θ – the **adjacent** side, to the hypotenuse.

Tan θ is the ratio $\frac{b}{a}$ of the opposite side to the adjacent side.

> **KEY FACT**   *sin θ = opposite/hypotenuse*
> *cos θ = adjacent/hypotenuse*
> *tan θ = opposite/adjacent.*

So, in *Fig. 1*, sin θ = b/a, cos θ = c/a and tan θ = b/c.

If you measure the lengths of b, c and a in *Fig. 1* you will find them to be about 12 mm, 20 mm and 24 mm respectively so that sin θ is approximately $\frac{12}{24}$ or 0.5 (which is exactly right for 30°). Cos $\theta = \frac{20}{24}$ or 0.83 is close to the correct value of 0.8660 and tan $\theta = \frac{12}{20}$ = 0.60 is close to the correct value of tan 30° which is 0.5773.

Trig ratios are used a great deal in A level physics. As an example, consider the equation that has to be satisfied if a diffraction grating is to send light of wavelength λ in a direction θ. It is $d \sin \theta = n\lambda$.

Example

$d \sin \theta = n\lambda$ for a *diffraction grating* where d is the distance between the lines of the grating, n is any small whole number and λ is the wavelength of light emerging at angle θ from the grating. If d is 0.50×10^{-5} m, n is 1 and θ is 6.8° what is the wavelength of the light.

Answer

$$d \sin \theta = n\lambda \text{ so that } \lambda = \frac{d \times \sin \theta}{n} = \frac{0.5 \times 10^{-5} \times \sin 6.8°}{1}$$

Now sin 6.8° can be found from a calculator and equals 0.1184.

$$\therefore \quad \lambda = 0.5 \times 10^{-5} \times 0.1184 = 5.9 \times 10^{-7} \text{ m}$$

Trig ratios equal to 1

If the angle θ in *Fig. 1* is made bigger then its sine value gets larger. The opposite side and hypotenuse finally become extremely large and equal when θ becomes 90°. So sin 90°= 1 and the sine of an angle can never exceed 1. This is worth remembering. Similarly if θ becomes very *small* then the adjacent side and hypotenuse become equal and cos θ = 1. The cosine of an angle can never exceed 1. There is (think about *Fig. 1*) no such limit to tan values.

> **KEY FACT**   *sin 90°= 1 and cos 0°= 1.*

Trig ratios of very small angles

If you are learning about lenses you may find it useful to know also that when any angle is very small its sine becomes very close to the size of the angle measured in radians. For example if $\theta = 3°$ which means $\theta = 0.0524$ radian then sin θ from your calculator is 0.0523. The reason for this is that when θ becomes very small a triangle containing it becomes indistinguishable from a sector of a circle so that length a or c in *Fig. 1* could be the radius of a circle and length b would be an arc. Then θ in radians = $\frac{b}{a}$ and equals sin θ. It is easy to show that, again for small angles only, tan $\theta = \theta$ in radians and cos θ becomes close to 1 (agreeing with the cos 0° = 1 result mentioned above).

> **KEY FACT** *For small angles sin θ = θ in radians, tan θ = θ in radians.*

Obtaining trig ratios from a calculator

For an angle *which is displayed as an answer* on the screen its sine can be obtained by pressing the key labelled 'sin' and then = .
If no entries have been made the sin key can be pressed followed by the number whose sine is wanted or the angle can be entered, the = key is pressed to make this angle the answer and then the sin and = keys are pressed. The procedure is similar for cos and tan values.
For sin 30° pressing the keys [SIN] [3] [0] [=] gives the answer 0.5
Using

[AC] [2] [0] [+] [1] [0] [=] [SIN] [=]

will give first 30 and then the 0.5 for the sine.

If you are working in degree mode and want the trig ratio for an angle that you've worked out in radians, a sine for example, then use the sin key, followed by entering the radian value, changing this angle to degrees using the shift key for DRG and selecting D and finally pressing the = key.

For the sine of 1 radian, starting with a cleared screen, a sine value of 0.84147 will be obtained by using the keys [SIN] [1] [SHIFT] [DRG] [R] [=]
The calculator will remain in degree mode.
If instead you press the sin key, change to radian mode using the mode key and enter your radian value you get the correct answer but are left in radian mode.

Test Yourself

Exercise 11.1.1

1 Find sin, cos and tan θ in *Fig. 1* for $a = 13$ cm, $b = 5$ cm and $c = 12$ cm.

2 Calculate tan θ for the smallest angle in the right-angled triangle having sides of 3, 4 and 5 cm.

3 The longest side of a triangle is a and $a^2 = b^2 + c^2 - 2bc \cos A$. If $b = 5$, $c = 6$ and $A = 50°$ what is the value of a?

4 Calculate the wavelength of light that produces a second order ($n = 2$) beam at 40° with a diffraction grating with grating element of 0.20×10^{-5} m.

> **HINT** *See the worked example on page 154.*

Inverse trig ratios

Inverse always means 'upside-down' or 'the opposite'. Here it means the opposite to finding the trigonometric ratio when you know the angle. So the inverse of finding sine of 30° which is 0.5 is entering of 0.5 and finding the angle is 30°.
The abbreviation for the inverse of sine is $\sin^{-1}$. (The −1 is used because −1 as an exponent of a number such as $\frac{2}{3}$ means the reciprocal of the number which means the number upside-down namely $\frac{3}{2}$. Here however it serves no purpose other than to mean the opposite of sine.)
In simple words $\mathbf{\sin^{-1} x}$ means 'the angle whose sine is x' written as $\sin^{-1} 0.5 = 30°$

KEY FACT *sin⁻¹ means 'the angle whose sin is'.*

On your calculator the key for inverse sine is marked with $\sin^{-1}$ as expected. It is the second function for the sin key. So the procedure for finding the angle whose sine is say 0.5 is

[SHIFT] [sin⁻¹] [0] [.] [5] [=] or [0] [.] [5] [=] [SHIFT] [sin⁻¹] [=]

Example

For a diffraction grating if light of wavelength (λ) equal to 500 nm is to fit the equation $d \sin \theta = n\lambda$ when d is 4×10^{-5} m and $n = 2$ what must be the value of θ?

Answer

$$\text{Sin } \theta = \frac{n\lambda}{d} = \frac{2 \times 500 \times 10^{-9}}{4 \times 10^{-5}} = 2.5 \times 10^{-2} = 0.025 \text{ and } \sin^{-1} 0.025 = 1.433° \text{ or } 1.4° \text{ 2 s.f.}$$

Example

When light enters a glass surface as in *Fig. 2* it is **refracted**. This means that the light changes direction.

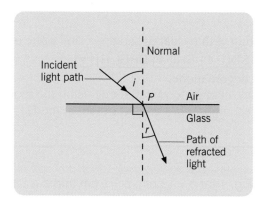

Fig. 2 *Refraction at an air-to-glass surface*

The ratio of $\frac{\sin i}{\sin r}$ is a constant called the **refractive index** and for a boundary between ordinary glass and air its value is 1.5. Calculate the angle of incidence needed to give an angle of refraction of 20° if the refractive index is 1.5.

Answer

$\frac{\sin i}{\sin r} = 1.5$ so that $\sin i = 1.5 \times \sin 20° = 1.5 \times 0.3420 = 0.5130$

$\therefore \quad i = \sin^{-1} 0.5130 = 30.87°$ or 31°

Test Yourself

Exercise 11.1.2

1 Obtain values of:

(a) $\sin^{-1} 0.60$ (b) the angle whose sine is 0.40

2 Obtain the value of

(a) $\cos^{-1} 0.9999$

HINT ▷ *Expect almost zero.*

(b) $\tan^{-1} 1.005$

HINT ⟩ *Expect answer close to 45°.*

3 If light arrives at an angle of 15° to the normal at an air-to-glass boundary for which the refractive index is 1.5, what is the expected angle of refraction?

HINT ⟩ *See the worked example above.*

Relationships between cosines and sines

When we want to write the square of a trigonometric ratio such as $\sin \theta$ we do not write $\sin \theta^2$ because this would be the sine of a squared angle. We could write $(\sin \theta)^2$ but it is simpler to write $\sin^2 \theta$.

KEY FACT *The square of $\sin \theta$ is $\sin^2 \theta$ and is called 'sine squared theta'.*

For any angle θ there is a relationship between $\cos^2 \theta$ and $\sin^2 \theta$ which is really a statement of the Pythagoras' formula. It is

KEY FACT $\cos^2 \theta + \sin^2 \theta = 1.$

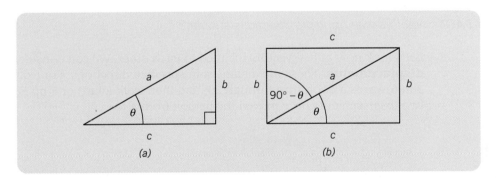

Fig. 3 *Relating cosines and sines.*

Fig. 3a shows a right-angled triangle for which $a^2 = b^2 + c^2$ (Pythagoras formula) and dividing both sides of the equation by a^2 gives $1 = \dfrac{b^2}{a^2} = + \dfrac{c^2}{a^2} = 1$

But $\dfrac{b^2}{a^2} = \sin^2 \theta$ and $\dfrac{c^2}{a^2} = \cos^2 \theta$ so that $\cos^2 \theta + \sin^2 \theta = 1$

KEY FACT $\cos^2 \theta + \sin^2 \theta = 1.$

This fact is used in Chapter 15 when simple harmonic motion is explained.
Another relationship involving sine and cosine tells us that:

KEY FACT $\cos(90° - \theta) = \sin \theta, \sin(90° - \theta) = \cos \theta.$

These facts are explained by *Fig. 3b* where you can see that $\sin \theta = \frac{b}{a}$ and also $\cos(90° - \theta)$ equals $\frac{b}{a}$. The relationship is useful when working with vectors later in this chapter. It means for example that $\cos 60° = \sin 30°$ and $\cos 30° = \sin 60°$.

11.2 Vectors

Vectors are quantities that have direction as well as size. An example is a force which may be directed downwards or in some other direction. An object may be moving and its velocity has direction. In contrast a mass, volume, area or energy cannot have direction and is a **scalar** quantity, not a vector. Forces are the most important vectors and working out the effect of a number of forces acting on a vehicle, part of a bridge or a person's leg can require the use of trigonometric ratios.

Combining vectors

If we have a force of 3 people pushing a car forward plus an additional force of 2 people pushing in the same direction the overall effect or **resultant** is a force of 5 people.

> **KEY FACT** *Vectors in the same direction simply add.*

Contributions that combine to form a vector are called its **components**. Of course if the additional 2 people pushed or pulled in the opposite direction the overall effect would be the same as having a single person pushing forwards.

> **KEY FACT** *Vectors with opposite directions subtract.*

Let's look now at two vectors inclined to each other. We can consider first two **displacements**. These are distances in specified directions. For example a person may walk 3 m towards a door, turn through 30° and then walk a further 2 m. *Fig. 4* shows how these two displacements have moved the person from A to C.

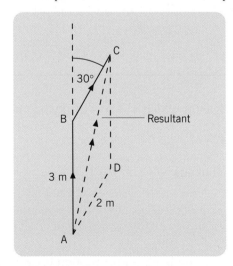

Fig. 4 *Combination of two displacements.*

> **KEY FACT** *The result of combining the two displacements is called the **resultant**. The resultant has the same effect alone as the original two vectors together.*

The directions of the displacements are shown by arrows and using a double arrow to show the direction of the resultant is a useful idea.

Fig. 3 has been carefully drawn to a scale of 1 cm for each metre of the real displacements. By measuring the distance AC and finding that it measures 4.8 cm we determine that the real displacement has a size (or **magnitude**) of 4.8 m. With the aid of a protractor the angle BAC is found to be 21°. So the resultant displacement is 4.8 m in the direction 21° from the initial displacement.

> **KEY FACT** *The resultant of two displacements can be found by scale drawing.*

A vector such as a force that contributes to a resultant, in other words is part of a larger force, is described as a **component** of that force.

So the 2 m and 3 m displacements were components of the 4.8 m displacement.

> **KEY FACT** *A resultant is formed from components.*

The parallelogram rule

Fig. 4 also illustrates the fact that the same resultant would be obtained if the 2 m displacement occurred first taking the person from A to D, followed by the 3 m displacement. It is possible too to imagine that the two displacements might occur at the same time. For example a person might move the 2 m across the floor of a vehicle that is at the same time making the movement of 3 m.

We realise that the order in which the displacements occur is not important. So it is quite usual for the two displacements to be represented by the parallelogram ABCD in *Fig. 4* rather than a triangle like ABC or ADC and the displacements are represented by the sides AB and AD so that the directions of the two displacements are both outward from A. The resultant is then also outward from A (*Fig. 5*).

The **parallelogram rule** can be applied not only to the combination of displacements but to any vectors. Consider next that the two displacements of 2 m and 3 m which we combined occurred at the same time and took 1 second. This means that we had a velocity of 2 m per second and a velocity of 3 m per second and the combined effect was 4.8 m covered in 1 second. Hence we conclude that the method enables velocities to be combined. Equally if we had two changes of velocity of 2 m s^{-2} and 3 m s^{-2} they would combine to give a resultant acceleration of 4.8 m s^{-2}.

Now this fact has an important consequence when accelerations are being combined. Suppose that the mass that is accelerated is 1 kg then the 2 m s^{-2} acceleration must be due to a force (= mass × acceleration) of 2 newton and the other acceleration must be due to 3 newton. The resultant acceleration we have seen is 4.8 m s^{-2} and must be due to a force of 4.8 N. Thus the parallelogram method and *Fig. 4* could be used for the 2 N and 3 N forces to give 4.8 N at 21° to the 3 N force (9 from the 2 N force).

> **KEY FACT** *The parallelogram method works for displacements, velocities, accelerations, forces and any other vectors.*

Fig. 5 shows how a vector parallelogram will usually look. It tends to be called a **parallelogram of forces** because it is mostly used for combining forces.

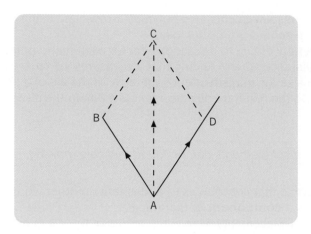

Fig. 5 *A vector parallelogram (or **parallelogram of forces**).*

KEY FACT   *The parallelogram rule states that the resultant of two vectors can be found by representing them in size and direction by two sides of a parallelogram and then the resultant is represented by the diagonal of the parallelogram, the directions of the three vectors being away from a common point.*

Example

An object having a mass of 1.0 kg rests on a very slippery horizontal surface. Two cords are attached to it and these pass over pulleys and have weights at their ends such that one cord pulls on the object with a 10 N force and the other with a 5.0 N force (see *Fig. 6(a)*).

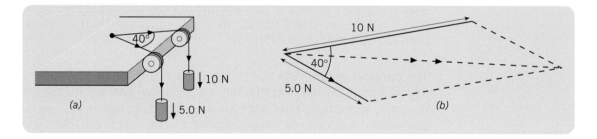

Fig. 6

The angle between the cords is 40°. The object is initially prevented from moving. When the object is released in what direction will it move and with what acceleration if no friction force or other retarding force is present?

Answer

The parallelogram of forces diagram is shown in *Fig. 6(b)*. (In practice the diagram would be made larger so that accurate measurements could be made from it.)
The scale used is 0.5 cm for each newton.

The measured length of the resultant is 7.0 cm so that the resultant force is 14 N and this force will produce an acceleration of 14 N/1 kg = 14 m s^{-2}. The angle between the resultant and the 10 N force is found to be 13°. The answer therefore is that the object will move off at 13° to the 10 N force, 27° to the 5.0 N force and the initial acceleration will be 14 m s^{-2}.

Forces at right angles

We continue this discussion of vectors on the assumption that we will be dealing with forces. Now suppose that two forces F_1 and F_2 are acting on an object and the forces happen to be at right-angles to each other. Combining here is an easier problem because we can avoid the need for a precisely-drawn scale diagram. The situation can be described by *Fig. 7(a)*. The vector diagram is now a rectangle instead of a parallelogram and is drawn in *Fig. 7(b)*.

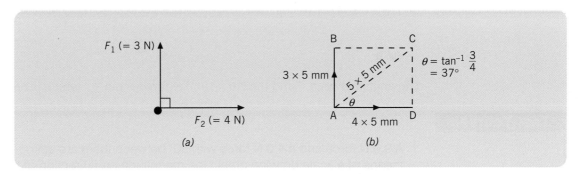

Fig. 7 *Forces at right-angles.*

We are able now to use the Pythagoras formula and calculate the length AC and no measurements using a ruler are needed.

$$AC^2 = AB^2 + BC^2$$
$$\therefore \quad AC^2 = 3^2 + 4^2 = 9 + 16 = 25$$
$$\therefore \quad AC = \sqrt{25} = 5$$

If the diagram had really been drawn to a scale of 5 mm for each newton and AC were then measured it would give AC = 25 mm but whether by scale drawing or by use of the Pythagoras formula the resultant force is 5 newton.

KEY FACT   *Forces at right-angles may be combined using the Pythagoras formula.*

The direction of the *resultant* can be found from the tangent value of θ which equals, in *Fig. 7(b)*, the ratio $\frac{CD}{AD}$, i.e. the ratio $\frac{3}{4}$ or $\frac{F_1}{F_2}$.

KEY FACT   *The resultant's direction is at angle $\tan^{-1}\frac{F_1}{F_2}$ to F_2.*

Example

An electron is subjected to a force of 2.0×10^{-15} N and at right-angles to this force a second force of 3.0×10^{-15} N. The gravitational force on the electron is negligible compared to these other forces. What is:

(a) the resultant force on the electron

(b) the angle between this force and the magnetic force?

Answer

Because the two forces are at right-angles the parallelogram of forces is a rectangle (*see Fig. 8*), the resultant force F is represented by a diagonal and the forces may be combined by use of Pythagoras's formula. Working in units of 10^{-15} N we have

$$F^2 = 2^2 + 3^2 = 4 + 9 = 13$$
$$\therefore \quad F = \sqrt{13} = 3.6 \text{ and } F = 3.6 \times 10^{-15} \text{ N}$$

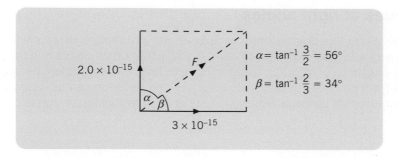

Fig. 8

The angle between F and the 2.0×10^{-15} N force is $\tan^{-1}\frac{3}{2}$ which is $56.3°$.

Test Yourself

1 A 9.0 N force and a 4.0 N force with 50° between them are acting on a small object. By means of a scale drawing obtain the magnitude and direction of the resultant.

2 Calculate the magnitude and direction of the resultant of 55 N and 25 N acting at right-angles on a small object.

3 A boat travelling forward at a speed of 3.0 m s^{-1} is being carried by a 1.0 m s^{-1} tide flowing at right-angles to the boat's forward direction. Calculate the magnitude and direction of the boat's resultant velocity.

11.3 Resolving forces

Splitting a force into two components

Later we will see some benefits of replacing one force by two forces but first consider this situation. A boat is to be towed along a canal with a force of 10 kilonewton directed forward, parallel to the canal bank so that minimum use need be made of the rudder. The force is to be provided by two ropes as shown in *Fig. 9(a)*. The two forces combine to give the required force and are therefore components of this force.

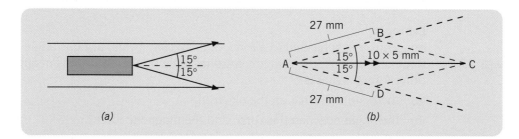

(a)   *(b)*

Fig. 9 *Use of two component forces.*

The procedure for determining the component forces required is the reverse of combining two forces into one resultant. It is called **resolving** the force. We can get the answers from a scale drawing. This is shown in *Fig. 9(b)* and the procedure for constructing the drawing is as follows.

First a line is drawn to represent the given force using an appropriate scale. A *scale* of 5 mm for each kilonewton has been chosen and a 5 cm line has been drawn (AC in *Fig. 9(b)*). Next the directions in which the components are required are marked in. These are dashed lines starting at A and making 15° angles with AC. Now lines (shown dotted in *Fig. 9(b)*) are drawn through C parallel to the dashed lines so that a parallelogram is completed. The points B and D are marked in and the sizes of AB and AD are measured. If the parallelogram were drawn using the number of centimetres suggested the size of AB and AD would be 27 mm showing that the required component forces are $\frac{27}{5} = 5.4$ kN from each rope.

KEY FACT   *A force can be resolved into two components using a parallelogram of forces.*

Resolving a force in two perpendicular directions

The word **resolving** here means finding components of the given force. We now consider the situation where the components are required to be, one of them at angle θ to the given force F and the other at right angles to the first component, as shown in *Fig. 10*.

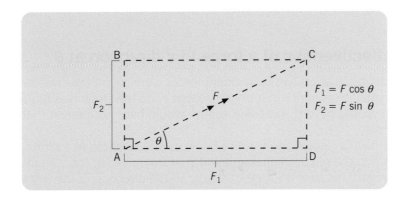

Fig. 10 *Resolving a force in perpendicular directions.*

The required directions are shown dashed. A parallelogram is completed and in this case is a rectangle and that is divided into two triangles. No scale is shown and the diagram can be of any size because, as we have found once before, a calculation can replace an accurate scale drawing if the forces concerned are represented by a rectangle. The required components are F_1 at angle θ and F_2 at right-angles to F_1 and these component forces can be calculated by use of trigonometry. $\sin \theta = \frac{F_2}{F}$ and $\cos \theta = \frac{F_1}{F}$ so $F_1 = F \cos \theta$ and $F_2 = F \sin \theta$.

KEY FACT   *A force can be resolved into components F sin θ and F cos θ at right-angles.*

Example
An object having a weight of 4.0 N is sliding down a 30° slope which is very slippery (zero friction force to hinder the sliding). With what force is the object pressed against the slope and with what force is it being accelerated down the slope?

Answer
We will replace the 4.0 N force (weight W) by two components that will together have the same effect as W. We choose the directions for these components to be perpendicular to the slope and down the slope as shown in *Fig. 11(a)*. The vector diagram is shown in *Fig. 11(b)*.

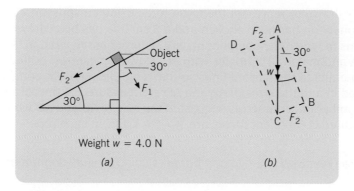

Fig. 11

In the triangle containing the 30° angle

$$F_1 = W \cos 30° = 4.0 \times 0.866 = 3.464 = 3.5 \text{ N and}$$
$$F_2 = W \sin 30° = 4.0 \times 0.5 = 2.0 \text{ N.}$$

The answers required are therefore that the force on the slope is 3.5 N and the accelerating force is 2.0 N.

The effectiveness of a force in a direction at θ

In the last calculation two component forces were considered instead of the given force. In that calculation suppose that only the force F_1 on the slope was required.
Then we would have $F_1 = W \cos \theta$ and we would have no interest in the component at right-angles to F_1.
So if we have a force F and want to know its effectiveness in a direction at angle θ, we think of $F \cos \theta$ in the required direction and an accompanying force $F \sin \theta$ (which is F_2) that we can disregard because it is perpendicular to the direction of interest and has no effect in that direction. We want $F \times$ cosine θ.

KEY FACT   *The effect of a force F in a direction at an angle θ to F is F cos θ.*

This book recommends that you learn the rule that the effectiveness of F at angle θ is $F \cos \theta$. If however you have a force F as in *Fig. 12* and an angle ϕ and want to find the effectiveness of F in a direction not at ϕ to F but at $90° - \phi$ you may want to use the ϕ angle because it will save your working out the angle θ that you would otherwise prefer to use. Or it may be that neither θ nor ϕ is known, you have introduced the symbol ϕ and don't want to introduce a second symbol – a second unknown quantity into your calculations (*see the example below*).

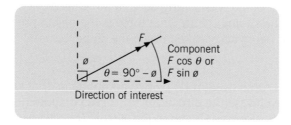

Fig. 12

KEY FACT   *Use F sin (angle) only when particularly convenient or essential.*

Example

A small object is fixed to the lower end of a cord which is inclined at 40° to the vertical. The tension force in the cord is sufficient to keep the object still (in equilibrium) in spite of it having a horizontal force of 0.40 N acting on it. Calculate the tension.

Answer

A suitable diagram is shown in *Fig. 13* where the tension is denoted by *T*.

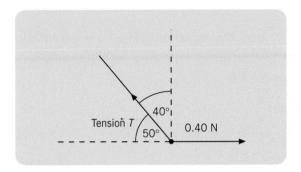

Fig. 13

The 0.40 N force to the right must be balanced by an equal force to the left. So the effect of *T* horizontally to the left must equal 0.40 N. But the effect of *T* in that direction which is at 50° to *T* (90° minus the 40°) is $T \times \cos 50°$ so that $T \cos 50° = 0.40$.

$$\therefore \quad T = \frac{0.40}{\cos 50°} = \frac{0.40}{0.6428} = 0.6223 \text{ or } T = 0.62 \text{ N}$$

(Note that the weight of the object, i.e. the pull of gravity, is pulling vertically down on the object but we did not need to involve this force. We were interested only in the horizontal forces.)

Example

A small object having a weight *W* of 2.0 N is supported by a string at an angle α to the vertical and also is pulled by a horizontal force *F* as shown in *Fig. 14*. If $F = 1.5$ N and the object is kept still, calculate the angle α and the size of the tension force *T* provided by the string.

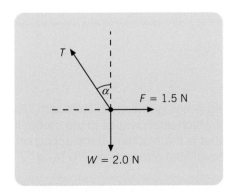

Fig. 14

Answer

Since the object is not moving in spite of the forces on it (we say that it is **in equilibrium**) the force W downwards (the object's weight) must be equalled by a force upwards i.e. by the vertical component of T.

$$W = T \cos \alpha$$
$$\therefore \quad 2.0 = T \cos \alpha \tag{1}$$

Also the force F to the right must be equalled by a force to the left. Thus

$$F = \text{horizontal component of } T \text{ to left}$$
$$F = T \sin \alpha \quad \text{(using } \sin \alpha \text{ to avoid introducing sine of another angle)}$$
$$\therefore \quad 1.5 = T \sin \alpha \tag{2}$$

To find α we can divide $T \sin \alpha$ from equation (2) by $T \cos \alpha$ from equation (1), and use the fact that $T \sin \alpha / T \cos \alpha = \sin \alpha / \cos \alpha = \tan \alpha$ (see earlier text) and hence deduce α.

$$T \sin \alpha = 1.5 \text{ and } T \cos \alpha = 2.0$$
$$\therefore \quad \tan \alpha = 1.5/2.0 = 0.75$$
$$\therefore \quad \alpha = \tan^{-1} 0.75 = 37°$$

Now to find T we can use either equation (1) or equation (2). From equation (1) we have

$$T = 2.0/\cos \alpha = 2.0/0.8 = 2.5 \text{ N}$$

Test Yourself

1 A heavy object is being dragged across a horizontal floor by a rope that makes an angle of 45° to the horizontal. If the tension is 200 N what forward force is the rope applying to the object?

> **HINT** *Sketches help most questions.*

2 The object shown in *Fig. 15* is in equilibrium under the action of the forces shown. W is the object's weight and T is the tension in each string. θ is the angle between each string and the vertical. Determine the value of T given that $\theta = 25°$ and $W = 12$ N.

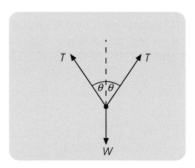

Fig. 15

3 An object having a mass of 20 kg is being pulled up a smooth incline at uniform acceleration by a force of 500 N which acts parallel to the plane. If the incline is at an angle of 20° to the horizontal, what is the resultant force acting on the body parallel to the plane? Assume the gravitational field strength, $g = 10$ N kg^{-1}.

> **HINT** *Sketch a labelled diagram. 500 N up. 200 cos 70°? or 200 sin 70° down?*

1 *Fig. 16* shows two equal masses supported by strings attached to two fixed points, A and D. String BC is horizontal. What is the approximate tension (a) in string AB and (b) in string BC? (Gravitational field strength, $g = 10$ N kg^{-1}.)

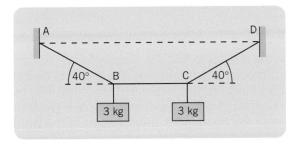

Fig. 16

HINT > *Find tension in inclined string by resolving vertically at B or C. Resolve horizontally to find other tension. 'Tension' means stretching force.*

2 In *Fig. 17* a person is supported by a parachute and is being pulled by a boat using a straight, lightweight line.
The weight of the parachute and person is 0.90×10^3 N and the tension in the line is 1.5×10^3 N. The line makes an angle of 30° with the horizontal.

(a) Calculate the horizontal component of the drag force on the parachute and person.
(b) Calculate the upward force on the parachute.

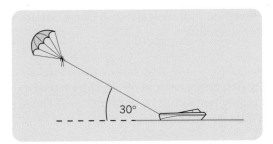

Fig. 17

HINT > *(a) Line makes 30° angle with horizontal at person. (b) Consider weight and component of tension.*

Resolving horizontal and vertical components

This *resolving* in horizontal and vertical directions as in the worked example on pages 165–6 should be regarded as a standard technique for determining forces on a body in equilibrium or their directions.

Really any two directions can be used but calculation is made easier if one or both directions coincide with forces acting on the body. Weight is always downwards so the vertical direction is usually a good choice and then the horizontal direction can be chosen to avoid having to further consider a component of the weight.

KEY FACT   *Resolving forces on a body vertically and horizontally is a standard technique.*

11.4 Equilibrium

Triangle of forces

The **triangle of forces** rule is used to find the sizes of forces or their directions when three forces acting on an object produce no movement. Their effects cancel. There is **equilibrium**. The parallelogram of forces has been recommended for combining two forces but just half a parallelogram, in fact a triangle, could be used instead. So in *Fig. 4* just triangle ABC could be used to find the resultant. In the figure AC is the resultant of the two forces combined. A force of equal size but oppositely directed (in direction C to A) would exactly balance the resultant or the two forces it replaced. So the triangle ABC with all three forces clockwise would indicate equilibrium. The triangle is the triangle of forces.

KEY FACT   *If three forces are in equilibrium they can be represented in size and direction by the three sides of a triangle with all arrows clockwise or all anticlockwise.*

The worked example below shows how the triangle of forces rule can be used. The triangle could be drawn carefully using the known values of forces and angles, with the triangle sides drawn exactly to scale, e.g. 1 mm for each newton of force, and with angles made exactly equal to those between the forces. The unknown forces or angles are then found by making measurements on the completed triangle.

Alternatively the triangle is sketched (not to scale) and trigonometry is used to calculate the answers. The worked example does this.

Example

An advertising sign is held by two cords each at 60° to the vertical as shown in *Fig. 18a*. The sign has a mass of 5.0 kg hanging beneath it by another cord. The sign has a negligible mass and is held in equilibrium. (Take $g = 10 \text{ m s}^{-2}$.)

(a)   Sketch a triangle of forces for the equilibrium.

(b)   Calculate the tension in each cord.

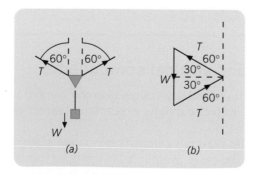

Fig. 18 (a) The advertising sign. (b) Triangle of forces.

Answer

(a) The triangle of forces shown in *Fig. 18(b)* is not drawn to scale. The weight W ($= 5.0 \times g = 50$ N) is drawn vertically downwards. (We draw forces as much as possible as they are seen.) The T sides of the triangle have to be drawn so as to complete the triangle with all arrows anticlockwise.

(b) Angles have been marked in the triangle of forces and weight $W = 2T \sin 30°$ so that

$$T = \frac{W}{2 \sin 30°} = \frac{50}{2 \times 0.5} = 50 \text{ N}$$

(Without the triangle of forces, part (b) would be answered by saying that each of the top cords provides an upward force of $T \cos 60$. So we have $2T \cos 60°$ upwards which equals 50 N downwards.)

Exam Questions

Exam type questions to test understanding of Chapter 11

Exercise 11.3.3

1 A ray of light travels from air to glass as shown in *Fig. 19*.

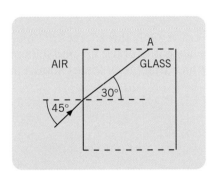

Fig. 19

The speed of light in air is 3.00×10^8 m s^{-1}. What is the speed of light in the glass?

A 2.00×10^8 m s^{-1} **B** 2.12×10^8 m s^{-1} **C** 3.00×10^8 m s^{-1} **D** 4.24×10^8 m s^{-1}

(OCR 2000)

> **HINT**
>
> *Refractive index of glass* $= \dfrac{\sin 45°}{\sin 30°} = \dfrac{\text{speed in air}}{\text{speed in glass}}$.

2 A diffraction grating has a spacing of 1.6×10^{-6} m. A beam of light is incident normally on the grating. The first order maximum makes an angle of 20° with the undeviated beam. What is the wavelength of the incident light?

A 210 nm   **B** 270 nm   **C** 420 nm   **D** 550 nm

(OCR 2000)

> **HINT**
>
> *Equation needed is $d \sin \theta = n\lambda$, with $n = 1$. See the worked example on page 154.*

3 (a) State the difference between scalar and vector quantities.
  (b) A lamp is suspended from two wires as shown in *Fig. 20*. The tension in each wire is 4.5 N.
      Calculate the magnitude of the resultant force exerted on the lamp by the wires.
  (c) What is the weight of the lamp? Explain your answer.

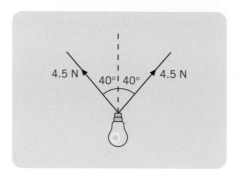

Fig. 20

(Edexcel 1999)

HINT ⟩ *F cos θ twice.*

4 *Fig. 21*(*a*) shows the jib of a tower crane. Only three forces act on the jib: the tension *T* provided by a supporting cable: the weight *W* of the jib; and a force *P* (not shown) acting at point X. The jib is in equilibrium. Which triangle of forces shown in *Fig. 22*(*b*) is correct?

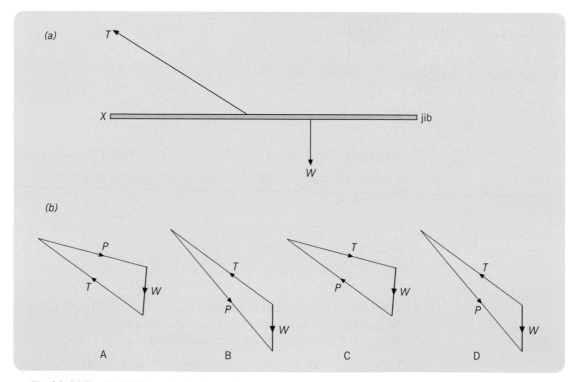

Fig. 21 *(a) The jib. (b) Triangles to choose from.*

(OCR 2000)

HINT ⟩ *See text for triangle of forces.*

Exercise 11.1.1
1 sine is 0.3846, cosine is 0.9231, tangent is 0.4167
2 0.75
3 4.74 cm
4 0.64×10^{-6} m

Exercise 11.1.2
1 (a) 36.9° (b) 23.6°
2 (a) 0.81° (b) 45.14°
3 9.9

Exercise 11.2.1
1 Approximately 12 N at 15° to the 9 N force

2 60 N at 24° to the 55 N force
3 3.16 m s^{-1}, 18.4° to the intended direction.

Exercise 11.3.1
1 141 N
2 6.6 N
3 0.43 kN

Exercise 11.3.2
1 47 N, 36 N.
2 (a) 1.3×10^3 N (b) 1.7×10^3 N

Chapter 12

Averages

After completing this chapter you should:

- *be able to define and calculate a mean value for a group of numbers*
- *have discovered how to use your calculator to obtain a mean value*
- *identify mean velocity with total distance divided by total time*
- *understand how a mean velocity equals $\dfrac{v_1 + v_2}{2}$ (the half-way value)*
- *know the meaning of root mean square (RMS).*

12.1 Mean values

A need for averages

It is often useful to describe a group of numbers by a single number that indicates in some way how big the numbers of the group are. The chosen number is then called the **average** value of the group. In physics the group of numbers may be a set of measurements of a certain quantity, the measurements being very similar but not exactly the same because of experimental errors. An average may be appropriate too, for describing the changing values of the speed of a car during a journey or a temperature during an experiment.

The **mean** value of a group of numbers or quantities is an average defined as the sum of all the values divided by how many values there are. For example the mean of the four numbers 3, 9, 5 and 7 is 6 because 3 + 9 + 5 + 7 equals 24 and this 24 divided by 4 gives 6.

> **KEY FACT** *The mean value = sum of numbers/how many numbers.*

You can use your *fx-83WA* calculator to obtain the mean of a set of numbers. Choose the statistics mode by using the mode key selecting SD by pressing 2. You need to clear the memory so press SHIFT, Scl and = (the Scl key is the same key as for AC). Numbers such as the 3, 9, 5, 7 just used can be entered, by pressing each number followed by the DT key (the same key as for M +). You get the mean by pressing SHIFT and the $\bar{x}$ (**x bar**) key (the same key as for 1), then =.

Ideally the term **average** would be used only in the way suggested so far in this chapter. The term *ordinary average* might be regarded as the same as *mean* while other averages would include for example the number that occurs most often in the group of numbers or the middle value of the numbers when placed in size order. However it is common for the single word **average** to be used as an alternative word for *mean* and in this book now there will be no difference between average and mean. This is true also for A level physics exams.

As explained in Chapter 7 there will always be some possible random errors associated with experimental measurements. *Fig. 1(a)* shows a set of eight results obtained for a certain quantity and *Fig. 1(b)* shows the distribution of the values obtained.

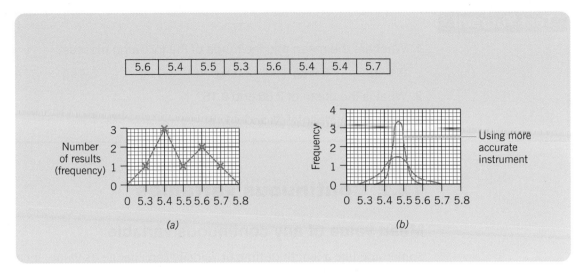

Fig. 1 *Experimental measurements.*

The points in the graph have been joined by straight lines because there are too few points for a smooth curve to be drawn.

If repeated measurements of the same quantity are not all the same but are showing random errors then a large number of these measurements, perhaps the times recorded for a certain movement, would produce a graph like the one shown in *Fig. 1(c)*. The results would be close to a smooth curve which is symmetrical about a central value. If in addition the results were obtained with only very small errors, perhaps by using a more accurate instrument, then the measurements would be mostly of the same size, say *x*, so that the graph would have a narrow peak at this value *x*. Clearly *x* is the true, error-free value.

KEY FACT *The peak of an error distribution curve occurs at the true size of the measured quantity.*

The mean value of a large number n of accurate measurements x is the total of values divided by n which is nx/n and therefore equals x. So the mean and true value are then the same.

KEY FACT *The mean of a large number of results = true value.*

The symmetry of the error curves is expected because for random errors there is as much chance of a result being above the true value as there is for a result to be equally below. As explained in Chapter 9 the drawing of a *best-fit* line between the points plotted for a graph is an averaging process.

KEY FACT *The mean value is the nearest we can get to the true value and is closer to the true value if more results are used.*

Range

For a group of results it may be desirable not only to know the mean but also to know how close the values are to the mean. The **range** of the results can then be stated. The range is the difference between the highest and lowest value.

Test Yourself

1 What are the mean and the range of the following masses?

| 2.1 kg | 3.9 kg | 2.0 kg | 3.0 kg | 2.7 kg | 4.3 kg |

2 What is the mean of 2.85 and 9.15?

3 What is the mean of $3x$ and $4.2x$?

12.2 Continuous variables

Mean value of any continuous variable

Something like a length or time or velocity can change its value. It can be a **variable** unlike π for example. Also it does not have to be a whole number. Its value can be between whole numbers. So it can be a **continuous variable**.

A velocity that increases with time is an example of a continuous variable for which we can calculate a mean value. For such a quantity we define the mean as follows. You will soon see that it fits in well with the definition of mean used so far.

> **KEY FACT** *For any change with time the mean = total of values for all time intervals divided by the number of the equal time intervals.*

If a velocity varies with time (as in *Fig. 2(a)*) from time v_1 at t_1 to v_2 at t_2 we can imagine this time to be divided into a large number n of equal small intervals each δt seconds as we did in Chapter 9.

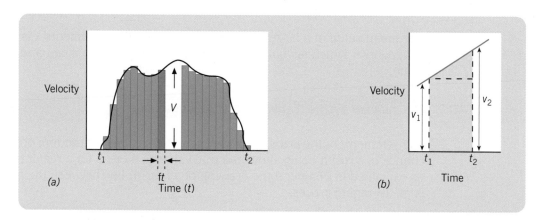

Fig. 2 *Velocity versus time graphs and areas beneath.*

The mean velocity for the whole time is the sum of all the velocities for the n time intervals divided by n. The Greek symbol Σ (pronounced *sigma*) is the abbreviation for 'the sum of all values of'. So mean velocity $= \Sigma v/n$ and this equals $\dfrac{\Sigma v \times \delta t}{n \times \delta t}$. But $v \times \delta t$ is the area under the graph for a time δt and so the sum of all such little areas gives the total area under the graph.

The product $n \times \delta t$ is the total time $t_2 - t_1$ and we conclude that mean velocity = area beneath graph divided by time concerned. Rewording this we have:

KEY FACT   *Area beneath graph = mean velocity × time.*

The area under the graph also tells us the total distance covered because each $v\delta t$ is a distance and the sum of these is the total distance. So

KEY FACT   *Distance covered = mean velocity × time.*

Comparing this equation with 'distance = velocity × time' for a constant velocity shows that a velocity which is changing can be regarded as constant if its mean value is used.

   A mean value allows a changing quantity to be regarded as constant.

Of course the variables need not be velocity and time. The electrical power being supplied to a heater could vary with time. The area beneath the power versus time graph would be the total energy supplied and a value given for mean power of an engine could be multiplied by time to get the energy used.

Example
A lake has a surface area of $2.0 \times 10^4 \text{ m}^2$ and a mean depth of 4.0 m. What is its volume?

Answer
We regard the lake as having a constant depth of 4.0 m.

The volume = area × depth = $2.0 \times 10^4 \times 4.0 = 8.0 \times 10^4 \text{ m}^3$.

Mean value for a steady change

Now suppose that we know that a velocity is increasing steadily with time (constant acceleration). The v versus t graph is as shown in *Fig. 2(b)*. The area under the graph (shaded) equals the area of the triangle plus that of the rectangle and works out to be $\frac{v_1 + v_2}{2} \times$ the time interval $(t_2 - t_1)$ and because the area is also mean v multiplied by the time we see that the mean velocity is $\frac{v_1 + v_2}{2}$ and this agrees with the definition of the mean for whole numbers.

   Mean of variable v as t changes $= \frac{v_1 + v_2}{2}$ if the changes in v and t are proportional.

Because the mean also equals $\frac{\text{total distance covered}}{\text{total time}}$ we get the important equation

KEY FACT   (mean velocity =) $\dfrac{s}{t} = \dfrac{u + v}{2}$. $\left(Mean = \dfrac{\text{initial value} + \text{final value}}{2}. \right)$

where s is distance covered, t time taken, u is initial and v final velocity. This is an important equation for physics calculations.

Example
A vehicle descends a slope with constant acceleration, its initial velocity being 1.0 m s^{-1}, its final velocity 5.0 m s^{-1} and the distance covered 8.0 m. Calculate the time taken.

Using $3y = x + 14$ first the expression for x is $3y - 14$ and substituting this in the equation $2y = 3x$ gives $2y = 3 \times (3y - 14)$ so $2y - 9y = -42$. Hence $-7y = -42$ meaning that $y = 6$. Putting this into the $2y = 3x$ equation gives $2 \times 6 = 3x$ so $x = \frac{12}{3} = 4$.

KEY FACT *Two simultaneous linear equations can be solved for x and y if an expression for x (or y) is obtained from one equation and this is substituted for y (or x) in the other equation to find x (or y).*

In equations where x and y are the only symbols used it is assumed that x may change and so cause change in y. So x and y are *variables*. Equations with other quantities as variables can also be solved, as the following calculation shows.

The equation $R = R_0(1 + \alpha\theta)$ relates a resistance R at temperature θ to resistance R_0 at $0\,°C$ and α is a constant decided by the material of which the resistor concerned is made. In a typical experiment the temperature is changed and this changes R while R_0 and α are not variable but constants. Now suppose that R_0 and α are unknown but measurements give us

$4.24 = R_0(1 + \alpha \times 18)$ and $4.8 = R_0(1 + \alpha \times 60)$

There are two unknowns (both constants) and two simultaneous equations.

From the first equation $R_0 = \dfrac{4.24}{1 + 18\alpha}$ and substituting for R_0 in the second equation gives

$4.8 = \dfrac{4.24}{1 + 18a} \times (1 + 60\alpha)$

$4.8(1 + 18\alpha) = 4.24(1 + 60\alpha)$

$4.8 + 86.4\alpha = 4.24 + 254.4\alpha$

$\alpha = \dfrac{4.8 - 4.24}{254.4 - 86.4} = \dfrac{0.56}{168} = 0.003333$ or 0.0033. The unit is $°C^{-1}$ or better still K^{-1}.

To find R_0 this value is used for α in the first equation which becomes

$4.24 = R_0(1 + 0.003333 \times 18)$ so that

$R_0 = \dfrac{4.24}{1 + 0.06} = \dfrac{4.24}{1.06} = 4.0$. The unit is Ω.

The method of solution described above will always work with *linear* equations. When we have simultaneous equations that are *not both linear*, solutions can be obtained using graphs, as explained later in this chapter. In the following *Test Yourself* exercise there are squared quantities u^2 and v^2 but the simultaneous equations can be solved easily. This is because u only occurs as a square and this can be regarded as a simple variable x throughout the calculation. Similarly v^2 can be regarded as y and the problem reduces to one with two linear equations.

The word **function** means a calculation process. As you know a calculator key may be used for two different functions by means of the $\boxed{\text{SHIFT}}$ key.

Saying 'y is a function of x' means that y is calculated by a process that involves x.

$y = 2 + x$ is an example. Our solving of simultaneous equations has been limited to linear functions.

Returning to our original problem of finding t from $a = \dfrac{v - u}{t}$ and $s = \dfrac{u + v}{2} \times t$ we had $v = 2 + 2t$ and $v = \dfrac{70}{t} - 2$. If we substitute $2 + 2t$ from the first equation into the second equation in place of v we get

$$70 = 2t + (2 + 2t)t \qquad \text{or} \qquad 70 = 4t + 2t^2 \qquad \text{or} \qquad t^2 + 2t - 35 = 0$$

and this is a quadratic equation. So our attempt to solve these equations for t leaves us with a problem of solving t when we have a quadratic equation.

Without using any numbers we can combine our two equations $a = \dfrac{v - u}{t}$ and $s = \dfrac{u + v}{2} \times t$ into one by the method used above. From the first equation we get $v = u + at$ and this is substituted into the second equation to give $s = \dfrac{u + u + at}{2} \times t$ which means $s = ut + \frac{1}{2}at^2$. If you know this equation it can be used to find t without using simultaneous equations but still you have to solve a quadratic equation to find t. Solving quadratic equations is explained below.

Similarly if you had the same equations $a = \dfrac{v - u}{t}$ and $s = \dfrac{u + v}{2} \times t$ and want a single equation without t in it you can get $t = \dfrac{v - u}{a}$ from the first equation and substitute this into the second equation in place of t there and get $s = \dfrac{u + v}{2} \times \dfrac{v - u}{a}$ which can be rearranged as $v^2 = u^2 + 2as$. So if your problem was to find v when t is unknown the solving of simultaneous equations is avoided.

Example

Two cars have speeds of 20 m s^{-1} and 30 m s^{-1} when a stop-watch is started at zero. The cars maintain steady accelerations of 4.0 and 1.0 m s^{-2} respectively. At what time will the first car have a speed twice that of the second car?

Answer

There is no reason why the cars should not be travelling in a straight line and the word **speed** can be replaced by **velocity**.

The relationship required is $v = u + at$ where v is velocity after time t, u is the initial velocity and a is the acceleration. Let the final velocities be $2x$ and x.

For the first car the final velocity is $2x = 20 + 4t$ and for the second car $x = 30 + t$.

From the first equation x is $10 + 2t$ and putting this into the second equation gives

$$10 + 2t = 30 + t$$
$$\therefore \quad t = 30 - 10 = 20 \qquad \text{The unit is second.}$$

| **Test Yourself** | Exercise 13.2.1 |

1 x and y fit both of the following two equations. What are the values of x and y?

$3y = 4x - 5$ and $2y = x + 5$

2 Solve the following pair of simultaneous equations for x and y.

$2y = 5x$ and $5y = 12x + 1$

3 A vehicle covered a distance of 50 m travelling with a constant acceleration a and reached a final velocity of 20 m s^{-1}. It was then tested over a distance of 100 m starting with the same initial velocity (call it u) and it reached a final velocity of 25 m s^{-1}. Assuming that its acceleration a, was the same in both cases determine u and a.

4 The impedance z of a circuit containing resistance R and capacitance C in series is given by $z^2 = R^2 + \left(\dfrac{1}{\omega C}\right)^2$ where ω denotes the angular frequency of the alternating supply potential difference and of the current in the circuit.

For a particular circuit z was 13 ohm when ω was 200 Hz and 23 ohm when ω was 50 Hz. Determine R and C. The SI unit for C is farad (F).

5 If $W = Fh$ where W is work done when force F acts over distance h and if $F = mg$ because F is the force due to gravity obtain a formula for W without F.

Using graphs to solve simultaneous equations

The pair of simultaneous equations $2x + 3y = 16$ and $5x - 2y = 2$ can be used as a simple example of solution by use of a graph. The first step in the procedure is to rearrange the equations so that each becomes a formula for one of the variables.

$$y = \frac{16 - 2x}{3} \text{ and } y = \frac{5x - 2}{2}$$

Next some values of y are found for one of the equations for a number of simple values of x. For the first equation when $x = 0$, y equals $\frac{16}{3} = 5.33$ and for $x = 1$, $y = \frac{14}{3} = 4.7$. A table of such results can be produced and a similar set of y values for various x is calculated for the second equation as in *Fig. 1*a.

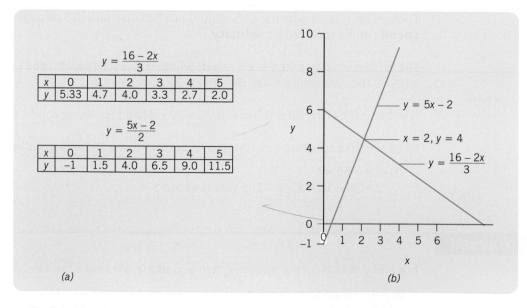

(a)

(b)

Fig. 1 *Solving simultaneous equations graphically.*

The tabulated x and corresponding y values are then used to plot a graph for each equation. As shown in *Fig. 1*b the graph lines cross (or **intersect**).

Now at any point on one line the y value there equals $\dfrac{16 - 2x}{3}$, i.e. the x and y values fit the first equation. Every point on the second line fits the second equation and at the point where the lines cross the x and y values fit both equations. From the figure the intersection occurs at $x = 2$, $y = 4$.

> **KEY FACT**   *At a point where two graph lines cross the x and y values fit the equations of both lines.*

The graph method of solving simultaneous equations will also work for graph lines that are curved, which happens when the simultaneous equations are not both linear as illustrated in *Fig. 2*.

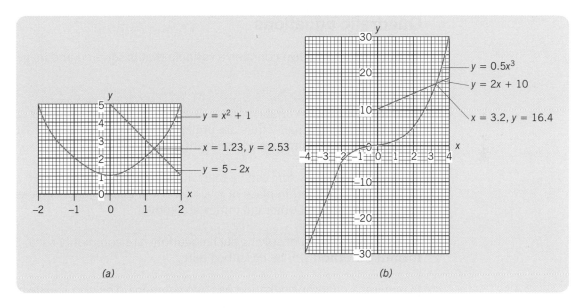

(a)   (b)

Fig. 2 *Solutions of simultaneous equations involving curves.*

It could be used for our finding t in the initial problem but we would like a quicker method.

Test Yourself

1 Plot a graph of $y = x^2$ and, using the same axes and scales, a graph of $y = 6 - x$ and hence obtain a solution for these equations regarding them as a pair of simultaneous equations.

> **HINT** >   *Make rough sketch of graph so as to choose suitable scales.*

2 A vehicle moves from rest with a constant acceleration of 4 m s^{-2}. Distances s_1 are recorded for times t as shown in the table below. A second vehicle starts off from rest and has the same accleration as the first but it starts 20 m ahead of the first vehicle's start and sets off 2.0 seconds later than the first vehicle. Use a graph method to determine the approximate distance travelled by the first vehicle before it catches up with the second vehicle.

Time t/s	0	1	2	3	4
First vehicle, distance s_1 ($= at^2/2$)	0	2	8	18	32
Second vehicle, distance s_2 ($= 20 + a(t-2)^2/2$)	–	–	20	22	28

> **HINT**
>
> *Plot s_1 and s_2 versus t on same graph (same scaling of axes). Make a rough sketch of the graph first.*

13.3 Solving quadratic equations

Quadratic equations

A **quadratic equation** contains a variable that is squared and its general form is

$$Ax^2 + Bx + C = 0$$

in which y and x are variables and A, B and C are constants. Lower case letters a, b and c could be used but small letters are preferred for variables, upper case for constants if the difference is to be emphasised. We want to determine x.

The equation $s = ut + \frac{1}{2}at^2$ can be rearranged to give $\frac{1}{2}at^2 + ut - s = 0$ and has the form $Ax^2 + Bx + C = 0$ with t in place of x, $A = \frac{1}{2}a$, $B = u$ and $C = -s$. Suppose that the time t is to be determined when all other quantities are known.

Three different ways of solving such equations are commonly used. Solving by use of the **formula method** will be described here.

In general a quadratic equation has two solutions. The very simple case of $x^2 = 4$ illustrates this fact because $x = +2$ and $x = -2$ both fit the equation. However the two solutions can be equal.

> **KEY FACT**   *A quadratic equation in general has two solutions. The two solutions may be the same.*

Some equations have no solutions. A simple example of an equation that cannot be solved is $x^2 = -1$. There is no number equal to $\sqrt{-1}$.

Solving a quadratic equation by the formula method

In general a **quadratic equation** consists of $Ax^2 + Bx + C = 0$ where A, B and C are constants, and so are not affected by the value of x if it changes. If there is a value of x that fits the equation we can find it, i.e. solve the equation for x, using the formula

$$x = \frac{-B \pm \sqrt{B^2 - 4AC}}{2A}$$

The formula is obtained as follows.

$$Ax^2 + Bx + C = 0$$

$$\therefore \quad x^2 + \frac{B}{A}x + \frac{C}{A} = 0$$

$$\therefore \quad x^2 + \frac{B}{A}x = -\frac{C}{A}$$

$$\therefore \quad \left(x + \frac{B}{2A}\right)^2 - \left(\frac{B}{2A}\right)^2 = -\frac{C}{A}$$

$$\therefore \quad \left(x + \frac{B}{2A}\right)^2 = -\frac{C}{A} + \left(\frac{B}{2A}\right)^2 = \frac{-4AC + B^2}{4A^2}$$

$$\therefore \quad x + \frac{B}{2A} = \pm \sqrt{\frac{B^2 - 4AC}{4A^2}} = \pm \frac{\sqrt{B^2 - 4AC}}{2A}$$

$$\text{and } x = \frac{-B}{2A} \pm \frac{\sqrt{B^2 - 4AC}}{2A}$$

KEY FACT   *If* $Ax^2 + Bx + C = 0$ *then* $x = \dfrac{-B \pm \sqrt{B^2 - 4AC}}{2A}$.

The square root of course has a + and a – value and to emphasise this the ± sign is shown in the formula for x.
Any quadratic equation that can be solved will be solvable by the formula method. You should learn or at least be able to recognise this formula.

As a simple illustration we can solve $2x^2 - 11x + 15 = 0$.

For this we have A = 2, B = –11 and C = 15 so that $x = \dfrac{-(-11) \pm \sqrt{121 - 4 \times 2 \times 15}}{2 \times 2}$

$$\therefore \quad x = \frac{11}{4} \pm \frac{\sqrt{1}}{4} = 2.75 \pm \frac{1}{4} = \text{either } 2.75 + 0.25 \text{ or } 2.75 - 0.25$$

$$\therefore \quad x = 3 \text{ or } x = 2.5$$

If a quadratic equation has no solutions the $B^2 - 4AC$ in the formula will be negative so that there is no possible value for the square root.

If $B^2 = 4AC$ then the square-root part of the formula is zero and there is a single answer for x (or you may say two equal answers for x).

So for obtaining t when we have $a = 2$, $u = 2$, $s = 35$ as suggested earlier in the equation $s = ut + \frac{1}{2}at^2$ we write $35 = 2t + t^2$ and $t^2 + 2t - 35 = 0$. So A = 1, B = 2 and C = –35 and the formula for t is

$$t = \frac{-2 \pm \sqrt{2^2 - 4 \times (-35)}}{2} = \frac{-2 \pm \sqrt{4 + 140}}{2} = \frac{-2 \pm 12}{2}$$

which means

$$t = -7 \text{ or } t = 5$$

Clearly 5 is the answer we want.

Solving simpler equations containing a squared unknown

Suppose a quadratic equation has no x term, e.g. $3x^2 = 11$. Well we simply get $x =$ either $\sqrt{\dfrac{11}{3}}$ or $\sqrt{3.667}$ which is ±1.915.

If instead a quadratic equation does not have the simple number term, for example $5x^2 = 6x$ then it can be written as $5x^2 - 6x = 0$ and this equals $x(5x - 6) = 0$. The conclusion is that either $x = 0$ or $5x - 6 = 0$ so $x = 0$ or $x = -\frac{5}{6} = -0.8333$. The two solutions are 0 and -0.8333.

Test Yourself

1 Solve the following equation (a) by factorising and (b) by use of the formula method:

$4x^2 + 9x + 2 = 0$

2 Solve each of the following quadratic equations by the formula method:

(a) $2x^2 - 0.15x - 0.05 = 0$ (b) $5x^2 + 2x - 9 = 0$

3 Solve each of the following equations.

(a) $5x^2 = 45$ (b) $5x^2 = 7$ (c) $4x^2 - 4 = 0$
(d) $2x^2 - 4x = 0$

HINT   *Cancelling of x values means x = 0 is a solution.*

(e) $7x^2 = 9x$ (f) $4x^2 = 9$

Physics examples

Solving a quadratic equation is not likely to be required in any A level physics exam but knowing about it could be useful in coursework, lectures or when reading a textbook. Three examples now follow. Another example of solving a quadratic equation has been given in Chapter 3 on page 42.

Example

In the formula $s = ut + \frac{1}{2}at^2$ if $s = 175$ m, $u = 10$ m s^{-1} and $a = 10$ m s^{-2}. What is the time t taken?

Answer

$s = ut + \frac{1}{2}at^2$

$\therefore \quad 175 = 10t + (0.5 \times 10t^2)$

$\therefore \quad 175 = 10t + 5t^2$

Rearranging this equation, $5t^2 + 10t - 175 = 0$.

Using the formula, $t = \dfrac{-B \pm \sqrt{B^2 - 4AC}}{2A}$

gives $x = \dfrac{-10 \pm \sqrt{10^2 + (4 \times 5 \times 175)}}{2 \times 5} = -1 \pm \dfrac{\sqrt{100 + 3500}}{10}$

$= -1 \pm \frac{60}{10} =$ either 5 second or -7 second.

The answer required is the time of 5.0 second.

Example

If an ordinary (**sinusoidal**) alternating current flows in a circuit containing inductance L henrys in series with capacitance C farads with negligible resistance the impedance z is given by the formula $z^2 = \left(\omega L - \dfrac{1}{\omega C}\right)^2$ where ω is the angular frequency or pulsatance of the current.

z is defined as the supply potential difference divided by the current (using RMS values or peak values – see Chapter 14).

If $L = 0.10$ H and $C = 1.0 \times 10^{-2}$ F what value or values of ω will give a current of 0.40 A RMS with a supply potential difference of 2.0 V RMS?

Answer

The equation given for z^2 can be simplified by taking the square root of both sides.

$$z = \pm \left(\omega L - \dfrac{1}{\omega C}\right)$$

The $\pm$ will first be regarded as $+$ and then the $-$ sign will be considered.

Using $z = \dfrac{V}{I}$, $\quad z = \dfrac{2}{0.4} = 5.0 \ \Omega$

Inserting the values in the z formula, $5 = (\omega \times 0.1) - \left(\dfrac{1}{\omega \times 10^{-2}}\right)$.

This equation can be simplified by multiplying all terms by $10^{-2}\omega$,

$5 \times 10^{-2}\omega = 10^{-3}\omega^2 - 1$

Multiplying all terms by 1000 and rearranging then gives $\omega^2 - 50\omega - 1000 = 0$ which has the form $A\omega^2 + B\omega + C = 0$ with $A = 1$, $B = -50$ and $C = -1000$.

Using the formula $\omega = \dfrac{-B \pm \sqrt{B^2 - 4AC}}{2A}$ gives

$$\omega = \dfrac{+50 \pm \sqrt{50^2 + 4000}}{2} = 25 \pm \dfrac{\sqrt{6500}}{2} = 25 \pm \dfrac{80.62}{2}$$

∴ either $\omega = 25 + 40.31$ or $25 - 40.31$, i.e. either 65.31 hertz or -15.31 hertz.

If the negative possibility of the $\pm$ sign is now considered for z the B becomes $+50$ and the answers for ω are $-25 - 40.31$ and $-25 + 40.31$, i.e. either -65.31 Hz or 15.31 Hz.

The answers required are 15.31 and 65.31 Hz, i.e. 15 or 65 Hz to 2 significant figures.

Example

An object is held at a distance u from a convex lens that has a focal length f, of 20 cm. The image is formed at a distance v such that $\dfrac{1}{u} + \dfrac{1}{v} = \dfrac{1}{f}$. If the distance between object and image is 100 cm what are the possible values for u and v?

Answer

From the $u + v = 100$ equation (this is the simpler equation) we have $u = 100 - v$.

Substituting $100 - v$ in place of u in the other equation gives $\dfrac{1}{100 - v} + \dfrac{1}{v} = \dfrac{1}{20}$.

Multiplying each term by $(100 - v) \times v$ gives $v + (100 - v) = \dfrac{(100 - v)\,v}{20}$ or $2000 = 100v - v^2$.

Rearranged this equation becomes $v^2 - 100v + 2000 = 0$ and this is a quadratic equation that can be solved by use of the formula $v = \dfrac{-B \pm \sqrt{B^2 - 4AC}}{2A}$ with A = 1, B = −100 and C = 2000.

$$\therefore \quad v = \frac{100 \pm \sqrt{10\,000 - 8000}}{2} = 50 \pm \frac{\sqrt{2000}}{2} = 50 \pm 22.36 = \text{either 72.36 cm or 27.64 cm}$$

(There are two suitable lens positions between the object and required image position, one in fact giving a small image and one an enlarged image.)

Test Yourself

1 Calculate the time required for a vehicle to cover a distance of 46 m starting with a velocity of 20 m s^{-1} and travelling with a constant acceleration of 3 m s^{-2}. ($s = ut + \dfrac{1}{2}at^2$)

2 For an object in simple harmonic motion (SHM) $a = -\omega^2 x$ where a is its acceleration away from the centre of the motion and ω is its angular frequency or pulsatance. A small platform is moving up and down in SHM with a maximum displacement of 0.050 m. To what value must the angular frequency be increased so that a small object resting on the platform will be left behind as the platform descends? (Take $g = 10$ m s^{-2} downwards.)

HINT ▷ *The required condition first occurs when the platform accelerates from its top position (maximum displacement) at a value equal to the acceleration due to gravity, $a = g$ towards centre, i.e. $a = -g$.*

Solutions of quadratic equations by use of a graph

Earlier in this chapter graphs were used to *solve simultaneous equations* and a similar technique can be used for equations such as *quadratic equations* which have only a single unknown quantity.

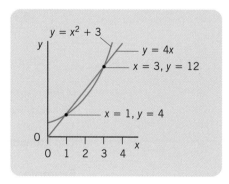

Consider $x^2 + 3 = 4x$. This is a quadratic equation for x and we expect two values of x to fit the equation. If a graph is plotted of $y = x^2 + 3$ and a graph of $y = 4x$ as in *Fig. 3* the two graph lines cross at two places. At such places $x^2 + 3$ and $4x$ are equal (both equal to y there) and so the x values at these places are the required solutions. In *Fig. 3* these solutions are seen to be $x = 1$ and $x = 3$.

Fig. 3 *Graphs plotted to solve $x^2 + 3 = 4x$.*

Exam Questions

Exam type questions to test understanding of Chapter 13

Exercise

1 A combination of x ohms and y ohms in series has a resistance of 60 ohms. A combination of two x ohm resistances and three y ohm resistances all in series has a resistance of 153 ohms. Deduce x and y.

HINT ▷ *Resistance of series combination (R) is given by $R = R_1 + R_2$. $x + y = 60$ and $2x + 3y = 153$.*

2 A ball is thrown upwards with an initial velocity of 15 m s^{-1}. Calculate the time it takes to reach a height of 10 m. (Take the acceleration due to gravity, g, as 10 m s^{-2}.)

> **HINT** *Distance $s = ut + \frac{1}{2}at^2$ where u = initial velocity, t the time and acceration $a = -g$.*

3 Two resistances x ohm and y ohm in series have a combined resistance of 250 ohm. In parallel their combined resistance is 60 ohm. Deduce x and y.

> **HINT** *For resistances in series see question 1. For parallel combination, $\dfrac{1}{R} = \dfrac{1}{R_1} + \dfrac{1}{R_2}$. Combine fractions as in Chapter 2.*

Answers to Test Yourself Questions

Exercise 13.2.1
1 $x = 5, y = 5$
2 $x = 2, y = 5$
3 $a = 2.25$ m s^{-2}, $u = 13.2$ m s^{-1}, ($u^2 = 175$)
4 $C = 1.0 \times 10^{-3}$ F, $R = 12\,\Omega$
5 $W = mgh$

Exercise 13.2.2
1 $x = 2, y = 4$
2 Theoretical answer is 24.5 m. Expect to get within 2 m of this figure.

Exercise 13.3.1
1 (a) $x = -\frac{1}{4}$ or -2
2 (a) 0.2 or -0.125 (b) -1.556 or 1.156
3 (a) $+3$ or -3 (b) ±1.183 (c) ±1 (d) 0 or 2 (e) 0 or 1.133 (f) ±1.5

Exercise 13.3.2
1 $t = 2$ s ($t = -\frac{46}{3}$ is impossible)
2 14 cycles per second or 14 Hz

Chapter 14

Logarithms

After completing this chapter you should:

- *know the meanings of ordinary and natural logarithms (log and ln)*
- *have used a calculator for logarithms*
- *be able to find antilogs*
- *be familiar with the properties of logarithms*
- *know when and how to use ln or log values for graphs such as for radioactive decay graphs.*

14.1 Logarithms and their properties

Introduction to logarithms

When a liquid escapes from the bottom of a tank the depth of liquid level falls fastest when the depth is greatest. Logarithms are useful for dealing with changes of this kind as you will see when logarithmic changes are considered on pages 200–1.

There are many ways of telling a person a number. Ninety-six can be written as 96 or 8 dozen or 24×4, the possibilities are endless. The **logarithm** (the **ordinary logarithm** or **log** or **Log**) of 96 is 1.982 and the **natural logarithm** of 96 is 4.564. If you are told that the log of a number is 1.982 then the number is definitely 96, the log is characteristic of that particular number. The relationship between a number and its logarithm is explained in the following section.

Ordinary logarithms (or common logarithms)

First, what is the ordinary logarithm or log of 10? Well, how many tens do you have to multiply together to make 100? The answer is 2 and this is the log of 100. For 1000 which equals $10 \times 10 \times 10$ there are three tens multiplying together. The log of 1000 is 3. Similarly the log of 100 000 (of 10^5) is 5 because the log of a number is the power of ten that is needed to equal that number.

So if you are told a number's log is 3 you know the number is 10^3 in the same way that three dozen tells us the number concerned is 12×3.

> **KEY FACT**   *number* $= 10^{\log \text{ of number}}$.

The 10 that is involved in these logarithms is called the **base**. So what have been called *ordinary logarithms* are logs that use 10 as their base.

> **KEY FACT**   *The logarithm to the base 10 of 100 000 or 10^5 is 5 and number $n = 10^{\log n}$.*

Natural logs

A **natural logarithm** uses as its base the very special number 2.718 which is more easily remembered by its name **e**. It is called the **exponential function**. This number e was mentioned in Chapter 4. Natural logs are met when you make radioactivity calculations. As an example the natural log of 9 is 2.197 meaning that $e^{2.197} = 9$. This is roughly what you would expect because $3^2 = 9$.

The log to base 10 of any number

Log 1000 = 3, log 100 = 2 and log 10 = 1 but what is the log of a number between one of these values. If a graph is plotted with a number x on the x-axis and log x on the y-axis the shape of the graph is as shown in *Fig. 1* and, in principle, we can say that the log of any number can be read from this graph.

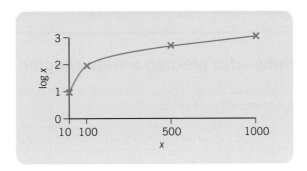

Fig. 1 *Graph of log of any number versus the number.*

We should find for example that the log of 500 is 2.7. The log of 96 mentioned earlier is obviously a little less than log 100, a little below 2.

Scientific calculators suitable for A level science work provide logs to base ten and natural logs of numbers.

KEY FACT   log *unless qualified is always taken to mean* log to the base ten.

On your calculator the key for log to base ten is called simply **log**. As an example try log 96 by entering

The value 1.982 should appear on the calculator's screen.

Now try finding log 500, answer about 2.7. Obtaining and using a number's log is often called *taking* its log.

Any number can be used as the base for logarithms but for A level physics we need only the logs mentioned so far. However it is useful to consider 2 as a base. For example log to the base 2 of 8 is 3 because 8 is $2 \times 2 \times 2 = 2^3$. To show the base being used for a log we use the base as a subscript to the abbreviation **log** so that $\log_2$ denotes **log to the base of 2** and $\log_2 8 = 3$.

If no base is specified it is assumed that the base is 10. In other words read $\log_{10}$ for log. For the so-called **natural** logarithms we write $\log_e$ or Ln or ln.

KEY FACT   *ln means the same as* $\log_e$.

To obtain a natural log on a calculator the procedure is the same as for an ordinary log but instead of using the log key the **ln** key is used. As an example try to find ln 9 and expect to get 2.197.

Test Yourself

Exercise 14.1.1

1 Without using a calculator write down the logs to base ten of:
   (a) 10 000   (b) 10   (c) 100   (d) a hundred thousand

2 Use a calculator to find the log of each of the following numbers, giving the value to 4 significant figures:
   (a) 100   (b) 1000   (c) 96   (d) 30
   (e) 3   (f) 300   (g) 3000   (h) 30 000

3 Use a calculator to find the natural log of each of the following numbers:
   (a) 10   (b) 2.75   (c) 100   (d) 148

A use for logarithms – for plotting a wide-ranging variable

The table below shows values of the potential difference across a particular type of diode and also the corresponding currents (I) obtained.

Potential difference/V	0.05	0.10	0.15	0.20	0.25	0.30
Current/microampere	1	4	20	100	400	1700

Fig. 2(a) shows the graph of these results with potential difference (V) on the x-axis and current (I) on the y-axis.

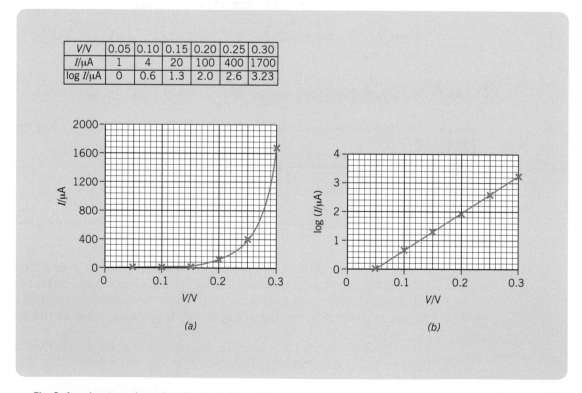

V/V	0.05	0.10	0.15	0.20	0.25	0.30
I/μA	1	4	20	100	400	1700
log I/μA	0	0.6	1.3	2.0	2.6	3.23

(a)

(b)

Fig. 2 *An advantage shown by a log scale for a wide-ranging variable. (a) Without using a log scale. (b) Using log I for y-scale*

A problem with this graph is that the lowest current values are indistinguishable from zero current. The graph only shows that the currents for potential differences of 0.05, 0.10 and 0.20 V are very small. It would be ridiculous to expand the y-scale to make all points plottable. (How big a page would be needed?) If instead we plot log current versus potential difference as in *Fig. 2(b)* then the points plotted for higher currents are brought closer together and all the results can be displayed clearly in a graph of reasonable size.

The log current values plotted are 0, 0.60, 1.30, 2.00, 2.60 and 3.23 respectively. You may wish to check these on your calculator.

Looking again at the graphs in *Fig. 2* you will see that the quantities plotted, as shown by the labelling of the axes, are pure numbers. In two of the graphs the ordinate is $I/\mu A$ and in the other graph the log is of course a pure number. Also the log must be the log of a pure number and this number is the current divided by its unit. For the x-axis the potential difference is divided by the volt and the label is V/V. As explained in Chapter 5 a label which is a quantity divided by its unit can include powers of ten without causing confusion and also it is appropriate for the graph scale figures to be shown without units.

Sometimes a log scale is used as suggested above but with the axis labelled with the original current values and not their log values. This is shown in *Fig. 3*.

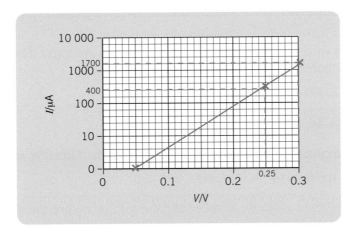

Fig. 3 *Use of a log scale without showing log values.*

The log scale in the case considered has been used with an ordinary **linear scale** on the other axis, a scale with evenly spread values (of potential difference). The linear scale represents potential difference values by proportional distances along the axis and *linear* implies this proportionality. But just think of linear scale as ordinary scale. You will also use graphs having log scales for both axes.

A feature that you must have noticed in the log graphs of *Fig. 2(b)* and *Fig. 3* are the straight lines obtained. This observation hints at other advantages of using logarithmic scales. These advantages are very important for A level physics and will be explained later in this chapter.

Antilogs

If you were given the graph of *Fig. 2(b)* and were asked what current is expected when the potential difference is 0.18 V you could read from the graph that log current is about 1.6. How then do you discover the current value from its log? You need the 'value whose log is', and this is the reverse of finding a logarithm. It is finding the **antilog**. The antilog in this example is 39.8 or approximately 40 microampere. You can confirm this by getting log of 39.8 on your calculator.

Now there is no key marked 'antilog' on your calculator. You use the 10^x key (the second function of the log key). Enter a log current (log I) value read from the graph (for instance 1.6) and 10^x is $10^{\log I}$ which equals I and is what you want. Try the key sequence

SHIFT 10^x 1 · 6 =   or   1 · 6 = SHIFT 10^x =

The antilog obtained is the $3.98 \times 10^1 = 39.8$ microampere.

As another example the antilog of 2 is 100 because log 100 = 2 and finding the antilog takes us from the 2 back to the 100. ($10^x = 10^2 = 100$.) Similarly the antilog of −2 is 10^{-2} or 1/100.

For natural logarithms antilogs are found on your calculator by use of the e^x key in place of the 10^x key.

KEY FACT *To find a number from its log use the 10^x key.*
To get a number from its natural log use the e^x key.

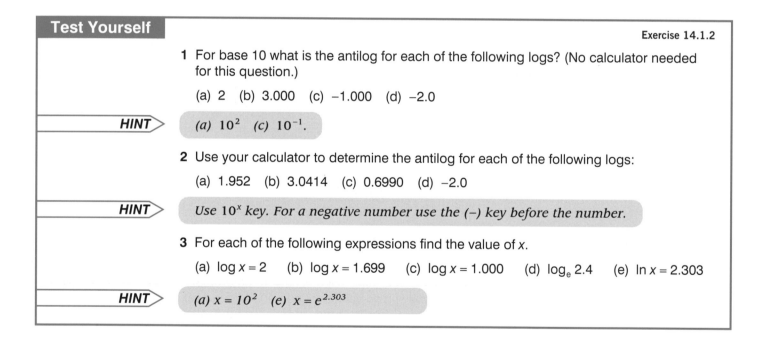

Test Yourself

Exercise 14.1.2

1 For base 10 what is the antilog for each of the following logs? (No calculator needed for this question.)

(a) 2 (b) 3.000 (c) −1.000 (d) −2.0

HINT *(a) 10^2 (c) 10^{-1}.*

2 Use your calculator to determine the antilog for each of the following logs:

(a) 1.952 (b) 3.0414 (c) 0.6990 (d) −2.0

HINT *Use 10^x key. For a negative number use the (−) key before the number.*

3 For each of the following expressions find the value of x.

(a) log $x = 2$ (b) log $x = 1.699$ (c) log $x = 1.000$ (d) $\log_e 2.4$ (e) ln $x = 2.303$

HINT *(a) $x = 10^2$ (e) $x = e^{2.303}$*

14.2 Rules concerning logarithms

There are several useful rules concerning logarithms that you need to know.

Can you simplify $10^{\log x}$? Well log x is always the power or exponent that must be placed above a 10 to get x. Here we have exactly that. We have log x above a ten. The expression simply equals x.

RULE 1 $10^{\log x} = x.$

The next rule tells us about the log of a product.

The log of a number is how many tens multiply to produce the number. For 100 000 this number is 5. For 100 it is 2 and for 1000 it is 3. Since 5 = 2 + 3 we can say

$$\log 100\,000 = \log(100 \times 1000) = \log 100 + \log 1000$$

RULE 2   *In general log ab = log a + log b.*

The rule can also be found by considering $10^{\log a} \times 10^{\log b}$ which, because of the definition of logarithms equals $a \times b = ab$. But by the same definition $ab = 10^{\log ab}$ so $10^{\log ab} = ab = 10^{\log a} \times 10^{\log b}$ and this, because indices add for a product, equals $10^{\log a + \log b}$.

Comparing the indices we see that log ab is the same as log a plus log b.

As an illustration log $(3 \times 7) = \log 21 = 1.3222$. Also log $3 = 0.4771$ and log $7 = 0.8451$ so that log $3 +$ log $7 = 0.4771 + 0.8451 = 1.3222$. Hence log$(3 \times 7) = $ log $3 +$ log 7.

Here is a rule concerning the log of a power, x^2 for example.

$$\log x^2 = \log(x \times x) = \log x + \log x = 2 \times \log x$$
So log $x^2 = 2 \times \log x$
Similarly for the general case

RULE 3   *log a^b = b log a.*

Now a rule concerning a reciprocal.

Log of $\frac{1}{a} = \log a^{-1}$ and using the relation log $a^b = b$ log a we get log $\frac{1}{a} = -$log a.

RULE 4   *log $\frac{1}{a}$ = log a^{-1} = $-$log a.*

Finally, what is the log of $\frac{a}{b}$? $\log \dfrac{a}{b} = \log\left(a \times \dfrac{1}{b}\right) = \log a + \log \dfrac{1}{b} = \log a - \log b$.

RULE 5   *log $\dfrac{a}{b}$ = log a $-$ log b.*

So far it has not been necessary to think about log of 1 or log of zero. Look at this sequence.

$$\log 1000 = 3 \quad \log 100 = 2 \quad \log 10 = 1 \quad \log 1 = ? \quad \log 0.1 = -1 \quad \log 0.01 = -2$$

You can guess that log 1 = 0, and this result is confirmed when log $\frac{a}{a}$ is written as log $a -$ log a (according to rule 5) and is seen to equal zero.

RULE 6   *log 1 = 0.*

As regards log 0 you can see from the sequence above that log 0.1 = -1 and log 0.01 = -2 and the log of any smaller value will be a larger negative number. Log of zero would be negative and extremely large.

$$\log 0 = -\text{infinity i.e. } \log 0 = -\infty$$

The ∞ symbol is the accepted symbol for **infinity** (it is an extremely large value).

Test Yourself

1 Simplify the following expressions.

(a) $10^{\log a}$ (b) $10^{\log 7}$ (c) $e^{\ln x}$ (d) $4e^{\ln 5}$

Test Yourself

1 Given that log 2 is 0.3010 what is the log of 200, i.e. of 2×100?

HINT ▷ $\log ab = \log a + \log b.$

2 Without using a calculator simplify:

(a) $y = \log 5 + \log 2$ (b) $\log a + \log b - \log c$

3 Rewrite each of the following expressions so that the factors are separated into different terms, for instance $\log ab = \log a + \log b$.

(a) $\log ax$ (b) $\text{Ln } by$ (c) $\log abc$ (d) $\log \frac{a}{b}$ (e) $\ln \dfrac{a}{bc}$
(f) $\log ab^2$

HINT ▷ *Remember* $\log b^2 = 2 \log b.$

(g) $\log ab^x$ (h) $\ln \dfrac{a}{b^y}$

When a variable is part of an exponent

Suppose that some experiment results were expected to obey the equation $y = a \times e^{bx}$ here x and y are variables. Suppose too that we don't know the values of the constants a and b but hope to discover these when we use measured values of x and y to plot a suitable graph. Note that x is part of the *exponent*.

Since $y = a \times e^{bx}$ we know that $\ln y = \ln$ of $(a \times e^{bx})$ and so $\ln y = \ln a + \ln e^{bx}$. Seeing that $\ln e^{bx} = b \ln e^x$ and so equals bx we get

$$\ln y = \ln a + bx \text{ or } \ln y = bx + \ln a.$$

This last equation is of the form $y = mx + c$ (or, to avoid using y for two purposes here we can say of the form *ordinate = a constant × absissca + another constant*). This means that a graph of $\ln y$ as ordinate plotted versus x as abscissa should give a straight line with gradient (m) equal to b and with an intercept on the ordinate axis of $\ln a$.

So taking logs of both sides of the equation has overcome the difficulty of having a variable in the exponent. Logs to base e (ln) are used because of the e in the equation. Using ordinary logs would not produce such a simple result.

As an illustration of this kind of graph you could have a voltage (V) which at any time t is given by $V = V_0 \, e^{-t/RC}$. V and t are the variables and R is known. What quantities would you plot to obtain a straight-line graph and determine C.

When time t is zero $V = V_0$ and subsequently V decreases.

Taking natural logs of both sides of the given equation we have

$$\ln V = \ln(V_0 e^{-t/RC}) = \ln V_0 + \ln e^{-t/RC} = \ln V_0 - \frac{t}{RC}$$

or

$$\ln V = \frac{-1}{RC}t + \ln V_0$$

This last equation has the form $y = mx + c$ which is the general equation for a straight-line graph. So plotting $\ln V$ as ordinate versus t as abscissa will give a straight line with gradient $m = \frac{-1}{RC}$. The gradient can then be measured and $C = \frac{-1}{Rm}$. The gradient is negative (i.e. the line is falling as t increases and m is negative) and consequently C works out as a positive answer. So a *logarithmic* scale is used for the *y*-axis, a *linear scale* for the *x*-axis.

KEY FACT *If $y = a\,e^{-bx}$ take logs to base e.*

The changing voltage in the above example is an **exponential change**. Such changes were discussed in Chapter 4. However if we take logs to base e of the equation $V = V_0\,e^{-\frac{t}{RC}}$ we get

$\ln V = \ln V_0 + \ln(e^{-\frac{t}{RC}})$. This is the same as $\ln V - \ln V_0 = e^{-\frac{t}{RC}}$ or $\ln\left(\frac{V}{V_0}\right) = \frac{-t}{RC}$ and means that

$t = -RC\ln\left(\frac{V}{V_0}\right)$. This shows a **logarithmic change** and means that an exponential change

and a logarithmic change are the same thing.

Example
The following data describe the variation of atmospheric pressure (P) with height (h) from the ground.

h/km	0	2	4	6	8	10	15	20
P/kPa	100	82	67	55	45	36	22	14
$\ln(P$/kPa$)$	4.6	4.4	4.2	4.0	3.8	3.6	3.1	2.6

Show that these results fit the relationship $P = P_0\,e^{-kh}$ or $\ln P = -kh + \ln P_0$ where P_0 is the pressure at height $h = 0$. Also determine the value of the constant k. (A graph of pressure versus h is plotted in *Fig. 4(a)* but this does not provide the answer.)

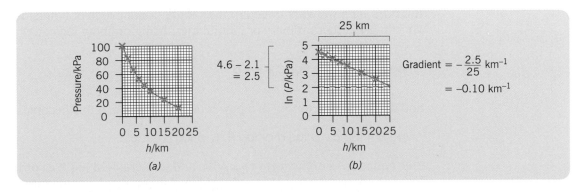

(a)   (b)

Fig. 4

Answer

Fig. 4b shows a straight line in agreement with the formula for ln (P). The slope of the graph is −k and equals −0.10 km⁻¹ meaning that $k = 0.10$ km⁻¹.

Test Yourself

Exercise 14.2.3

1 A circuit contains fixed values of inductance L and resistance R and the supply potential difference causes a current I_0 to flow. The potential difference is then suddenly reduced to zero. The formula for the decay of current I is $I = I_0\, e^{-t/T}$. T denotes the **time-constant** of the circuit and equals $\frac{L}{R}$. The value of R is known. What quantities would you plot to obtain a straight-line graph and how would you use the graph to discover L?

HINT $\ln I = \ln I_0 - \frac{t}{T}$.

2 Water is flowing from the bottom of a tank. The height of the water in the tank is h, and decreases with time t starting at height h_0 at time $t = 0$. h changes according to the formula $h = h_0\, e^{-ct}$, c is a constant. Which of the following combinations is correct for a graph that will be a straight line?

	y-axis	x-axis	gradient
A	ln h	t	c
B	ln t	h	$-c$
C	ln(h/h_0)	t	$-c$
D	ln h	t	h_0

HINT $\ln h = \ln h_0 - ct$ or $\ln \dfrac{h}{h_0} = -ct$.

Taking logs when a variable is to an unknown power

Consider the count-rate R for a nuclear radiation measured at a distance d from the source.

We may want to find out whether R agrees with the formula $R = \dfrac{c}{d^p}$ where c is a constant.

If it does then we want to determine p and c. Taking logs of both sides of the formula for R. We can use either ordinary or natural logs. Using ordinary logs gives

$$\log R = \log\left(\frac{c}{d^p}\right) = \log c - \log d^p$$

and using the rule for log of a power ($\log a^b = b \times \log a$) this becomes

$$\log R = \log c - p \log d \quad \text{or} \quad \log R = -p \log d + \log c$$

This equation is of the form $y = mx + c$ and so a graph of log R as ordinate plotted versus log d as abscissa will show a straight line with gradient $-p$ and intercept on the log R-axis of log c. The graph in *Fig. 5* is an example of this kind of graph.

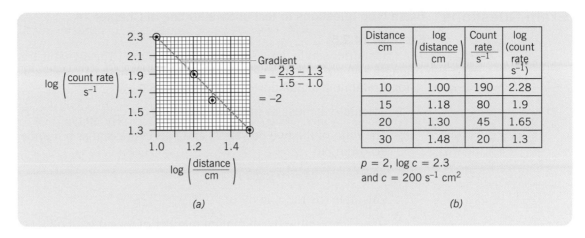

Fig. 5 *Plotting log count rate versus log distance.*

If it were suspected that $R = c/d^2$ a graph of R versus $1/d^2$ could have been tried and a straight line would confirm this relationship. However, if the line were not straight it would tell us very little, not allowing the value of p to be found so the log versus log graph is preferred. Note that logs are used on both axes (a *log log graph*) whereas for a variable that is an exponent only one axis used logs.

KEY FACT *To determine the power of a variable a log versus log graph may be used.*

For $y = x^p$, $\log y = p \log x$ and for $y = k \log x^p$, $\log y = \log k + p \log x$ and p is the gradient of the log log graph.

Test Yourself

Exercise 14.2.4

1 If a beam is supported at each end so that it is horizontal and then a weight is hung from the middle of the beam there will be a depression d at the middle of the beam such that $d = kL^p$ where L is the length of the beam between its supports and k is a constant for the beam. (k is affected for example by the beam's thickness.) Corresponding values of L and d are recorded using the same weight at the centre for all measurements. Suggest how a graph could be plotted and used to determine the exponent p.

2 For a simple pendulum the period is expected to be given by $T = 2\pi\sqrt{\dfrac{L}{g}}$. L and T are variables.

(a) Which of the following combinations for x and y-axes of a graph should give a straight line graph?

x	y
L	T
L	T^2
$\log T$	$\log L$
L	$\log T$

(b) Write down an expression for the gradient for each of the combinations listed in your answer to (a).

HINT $T^2 = \dfrac{4\pi^2 L}{g}$ *and so* $2 \log T = \log \dfrac{4\pi^2}{g} + \log L.$

Exam Questions

Q

Exam type questions to test understanding of Chapter 14

Exercise 14.2.5

1 A transmitter sends out a signal of power 60 mW. The signal is transmitted along a cable of length 140 km having an attenuation per unit length of 2.3 dB km^{-1}. The receiver operates for a minimum signal power at its input of 6.0 μW.

   (a) Calculate the loss in signal power along the cable.

   (b) The ratio of the two powers P_1 and P_2 is expressed as a number of decibels (dB) according to

   number of dB = $10 \log(\frac{P_1}{P_2})$

   Calculate the ratio, in dB of $\frac{\text{minimum power to receiver}}{\text{transmitted power}}$

   (c) Hence determine the minimum number of repeater amplifiers, each of gain 75 dB, which must be situated along the cable so that the received signal power is sufficient.

   (OCR 2000, part question)

HINTS AND TIPS

Give answer to part (a) in decibels (dB). Attenuation means reduction. Multiply dB per km by number of km (b)–(c) Shortfall of 322–40, i.e. 282 needed.

2 The activity (A) of a radioactive source decays according to the equation $A = A_0 e^{-\lambda t}$ where t is time.

   (a) Show that the time (T) required for the activity to fall to $\frac{A_0}{2}$ is given by $T = -\dfrac{\ln 0.5}{\lambda}$ or $\dfrac{0.693}{\lambda}$

   (b) (i) e can be written as 2^p. What is the value of p?

   (ii) Using 2^p in place of e show that the decay equation can be written as $A = A_0 2^{-\frac{t}{T}}$ or $A = \dfrac{A_0}{2^{t/T}}$

HINTS AND TIPS

(a) Take logs to base e. (b)(ii) replace λ by $\dfrac{0.693}{T}$.

3 Caesium-137 is a radioactive isotope. Its half-life is 8.5×10^8 s (27 years). A sample of caesium-137 is found to have an activity of 1.6×10^9 Bq.

   (a) Calculate the decay constant of caesium-137.

   (b) Calculate the number of moles of caesium-137 in the sample.

   (c) Calculate the activity of the sample after a time of 540 years.

   (WJEC 2000)

HINTS AND TIPS

A mole is 6.0×10^{23} particles (caesium atoms). Bq denotes decay of one atom per second. For a formula see question 2. Also $A = -\lambda \times$ number of atoms.
1 mole of atoms is Avogadro's number of them. Use λ in year^{-1} and T in years.

4 (a) For an adiabatic compression of a gas $P_1 V_1^{\gamma} = P_2 V_2^{\gamma}$ where γ is a constant for the gas concerned. γ can be determined by an experiment in which the pressures and volumes are measured. Obtain a formula for γ in terms of $\log P_1$, $\log V_1$, $\log P_2$ and $\log V_2$.

   (b) Use your formula to calculate γ for nitrogen gas from the following data.
   $p_1 = 1.1 \times 10^5$ Pa, $V_1 = 17 \times 10^{-4}$ m^3, $P_2 = 1.9 \times 10^5$ Pa, $V_2 = 11.6 \times 10^{-4}$ m.

HINT

Two minus signs are the same as +.

Answers to Test Yourself Questions

Exercise 14.1.1

1 (a) 4 (b) 1 (c) 2 (d) 5
2 (a) 2 (b) 3 (c) 1.982 (d) 1.477 (e) 0.4771
(f) 2.477 (g) 3.477 (h) 4.477
3 (a) 2.303 (b) 1.012 (c) 4.605 (d) 4.997
(e) −1.000 (f) −2.303

Exercise 14.1.2

1 (a) 100 (b) 1000 ($10^{3.000}$) (c) 0.1 (d) 0.01
2 (a) 89.54 (b) 1100 (c) 5 (d) 0.01
3 (a) 100 (b) 50 (c) 10 (d) 11.02 (e) 10

Exercise 14.2.1

1 (a) a (b) 7 (c) x (d) 20

Exercise 14.2.2

1 2.3010

2 (a) log 10 or just 1 (b) $\log \frac{ab}{c}$
3 (a) $\log a + \log x$ (b) $\ln b + \ln y$
(c) $\log a + \log b + \log c$
(d) $\log a - \log b$ (e) $\ln a - \ln b - \ln c$
(f) $\log a + 2 \log b$ (g) $\log a + x \log b$
(h) $\ln a - y \ln b$

Exercise 14.2.3

1 $\ln I$ versus t. $L = -R/\text{gradient}$
2 C

Exercise 14.2.4

1 Plot log d versus log L and gradient is p. (The logs can have any base, e.g. ln d versus ln L.)
2 L versus T^2 or log T versus log L

Chapter 15

Maths of circular motion, oscillations and waves

After completing this chapter you should:

- *be able to use radian measure for circular motion calculations*
- *appreciate the role of the radian and trigonometric ratios in the mathematics of simple harmonic motion*
- *understand the meaning of trigonometric ratios for angles greater than 90°*
- *understand equations describing waves*
- *understand what phasors are.*

15.1 Circular motion

Movement along a circular path

Examples of such movements were described in Chapter 10 and the terms **speed v**, **angular velocity ω**, **frequency f** and **period T** were defined. Important formulae obtained were

Angular velocity ω = angle turned through/time taken or $\omega = \frac{\theta}{t}$

$v = \frac{\text{arc}}{\text{time}} = \frac{r\theta}{t} = r\omega$

speed $v = \frac{2\pi r}{t}$ $\omega = \frac{2\pi}{T}$ and $T = \frac{2\pi}{\omega}$

frequency $f = \frac{1}{T}$

θ in these formulae denotes the angle turned through in radians, the **radian** having been explained in Chapter 10 on page 149.

Test Yourself

Exercise 15.1.1

1 What angular velocity corresponds to a frequency of 50 hertz? (Take $\pi = 3.14$)

2 A cyclist completes one lap of a circular track of 90 m radius in a time of 40 s. Calculate the average speed (a) in m per s and (b) in km per hour.

HINT   *See Chapter 5 for conversion of units.*

Maths of circular motion

The maths of **circular motion** makes use of vector combination which was explained in Chapter 11. It allows calculations to be made on the circular orbit of a satellite for example.

In *Fig. 1(a)* a small object is moving at constant speed v, angular velocity ω, along a circular path of radius r. The object could be a mass being whirled round at the end of a string.

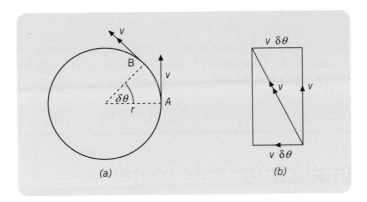

Fig. 1 *Motion along a circular path.*

The object passes through position A and reaches B a small time δt later. The small angle moved through is denoted by $\delta\theta$ and is measured in radians. The directions of the velocities at A and later at B are indicated by arrows in the diagram and the angle between these is easily seen to equal the angle $\delta\theta$.

A vector parallelogram for this movement is drawn in *Fig. 1(b)*. The new velocity is represented by the diagonal of the parallelogram and must be the resultant of the initial velocity v and the velocity represented by a short side of the parallelogram. Because $\delta\theta$ is very small the parallelogram must be imagined to be much thinner than shown and so the short sides are effectively perpendicular to the other sides and each short side has a length $v \times \delta\theta$ just as if it were a small arc of a circle with radius v.

The direction of this vector is perpendicular to the v direction and so is perpendicular to the circular path and directed towards the circle's centre. This added velocity $v\delta\theta$ is achieved in time δt, i.e. there is an acceleration, a equal to $v\delta\theta/\delta t$ or $v\omega$ or, since $\omega = \frac{v}{r}$, $a = \frac{v^2}{r}$.

KEY FACT *A steady movement along a circular path requires an inwards acceleration $a = \frac{v^2}{r}$.*

To produce this necessary acceleration there must be an inwards force F, perhaps provided by a string tied between the object and the circle's centre, and F must equal the object's mass × the acceleration a. Thus $F = \frac{mv^2}{r}$.

KEY FACT *For circular motion the required inwards force is $F = \frac{mv^2}{r}$.*

Example
An Earth satellite is orbiting the earth in an approximately circular orbit whose radius may be taken as 6000 km. Calculate the steady speed of movement along the orbit. (Take $g = 10 \text{ m s}^{-2}$.)

Answer
A force $\frac{mv^2}{r}$ must be acting on the satellite and directed towards the centre of the orbit which is the centre of the Earth. This force is the gravitational force of the Earth pulling on the satellite, i.e. the weight w of the satellite. $w = mg$ and using this fact,

$mg = mv^2/r$ (and the m on both sides cancels)

$\therefore \quad v^2 = gr = 10 \times 6000 \times 10^3 = 6 \times 10^7 = 60 \times 10^6$

$\therefore \quad v = \sqrt{60} \times \sqrt{10^6} = 7.746 \times 10^3 \text{ m s}^{-1} = 7.7 \text{ km s}^{-1}$

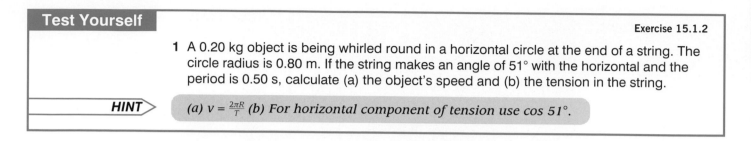

1 A 0.20 kg object is being whirled round in a horizontal circle at the end of a string. The circle radius is 0.80 m. If the string makes an angle of 51° with the horizontal and the period is 0.50 s, calculate (a) the object's speed and (b) the tension in the string.

HINT   *(a) $v = \frac{2\pi R}{T}$ (b) For horizontal component of tension use cos 51°.*

15.2 Simple harmonic motions

The shadow of a rotating object

A **simple harmonic motion** (SHM) is a particular kind of to-and-fro movement (**oscillation**) that obeys the equation $a = -\omega^2 y$. This equation is explained below. Vibrations and the swinging of a pendulum are examples of this kind of movement. The SHM of a shadow of a rotating object is worth studying as a means of deriving the equations you need to know. Although obtained for the shadow these equations will apply to all other SHMs.

A small object (P in *Fig. 2*) is fitted to the circumference of a disc. The disc is rotated at a steady angular velocity ω and a shadow of P is produced on a screen at S by sending light past P as shown. The shadow moves up and down between T and B and M is the middle point of this movement.

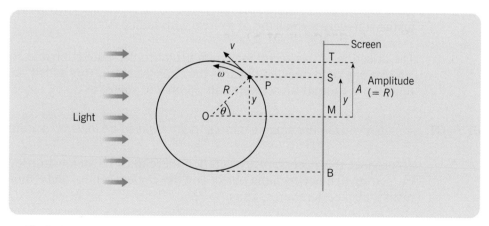

Fig. 2 *A simple harmonic motion.*

The greatest distance moved from the middle is called the **amplitude** (A) and equals MT or MB in *Fig. 2* and this amplitude equals the radius R of the rotating disc.

If a clock is started at a moment when the shadow is at M then angle $\theta = 0$ when time $t = 0$ and at other times $\theta = \omega t$. The distance y of the shadow from M is equal to $R \sin \theta$ or $A \sin \theta$ or $A \sin \omega t$. Note here that the ωt is not enclosed in brackets because it assumes that **sin** applies to the product. Brackets are used for the sine of a sum such as $\sin(\omega t + 5)$.

KEY FACT   *For SHM $y = A \sin \omega t$.*

The velocity of P is given by $v = R\omega$ (as stated earlier in this chapter) so that the velocity of S, which will be vertical, is $R\omega \cos \theta$ or $A\omega \cos \theta$ or $A\omega \cos \omega t$.

KEY FACT   For SHM $v = A\omega \cos \omega t$.

The acceleration of P as it moves in its circular path is v^2/R towards the centre O of the rotation and so the acceleration of S is $\dfrac{v^2 \sin \theta}{R}$ downwards in *Fig. 2* or $\dfrac{-v^2 \sin \theta}{R}$ in our positive direction (upwards). Using the fact that $v = R\omega$ for any circular motion the acceleration of S becomes $a = -R\omega^2 \sin \theta$ or $-A\omega^2 \sin \omega t$.

KEY FACT   For S.H.M $a = -A\omega^2 \sin \omega t$.

The above equation for a can be rewritten using the fact that $y = A \sin \omega t$ and we get $a = -\omega^2 y$ with the minus sign introduced to emphasise that when y is upwards (+ direction) a is downwards (negative). This is the equation used to define SHM. It shows that the acceleration a is proportional to the displacement y. The shadow movement is a simple harmonic motion.

Also the equation for v can be changed by using $A \sin \omega t$ for y. First we use the fact that $\sin^2 \theta + \cos^2 \theta = 1$ (*see Chapter 11*) so that $v = A\omega \cos \omega t = A\omega\sqrt{\cos^2 \omega t} = A\omega\sqrt{1 - \sin^2 \omega t} = \omega\sqrt{A^2 - A^2 \sin^2 \omega t}$ and then $v = \omega\sqrt{A^2 - y^2}$.

KEY FACT   For SHM $a = -\omega^2 y$ and $v = \omega\sqrt{A^2 - y^2}$.

Other examples of SHM

All the above equations have been derived for the shadow (projection of a circular motion) on a screen but if any to-and-fro motion of an object obeys any of these equations, for instance $a = -\omega^2 y$ then it moves like the shadow and is a SHM. The up and down movement of a mass attached to the bottom of a spring is an SHM.

For the spring and for other examples of SHM the quantity ω seems inappropriate, being the angular velocity of a circular motion. We don't see any rotation when a mass moves up and down on a spring! We just regard ω as an alternative to writing $2\pi f$ (see the start of this chapter) and describe it as the **angular frequency** of the motion rather than as the **angular velocity** which related to circular motion.

KEY FACT   For an SHM ω is called the angular frequency.

Note that the equations for SHM have been derived by considering the rise of a shadow and you might ask whether they still hold for the subsequent downward and upward movements and indeed for further cycles of the motion. They do and this will be explained on page 210.

Example
The bob of a simple pendulum is executing a horizontal simple harmonic motion having a period of 2.0 s and an amplitude of 0.15 m. Calculate:

(a)   the speed of the bob when it is passing through the lowest point C of its swing

(b)   the acceleration and speed of the bob when it is at 0.10 m from C.

(c)   How long does it take for the bob to travel from C to a displacement of 0.10 m?

Answer

(a) $v = \omega\sqrt{A^2 - y^2}$ and at C $y = 0$ $\therefore$ $v = \pm\omega A$

Since the period $t = 2.0$ s the frequency $f = \frac{1}{2} = 0.5$ Hz, $\omega = 2\pi \times 0.5 = \pi$

$\therefore$ $v = \pm\pi \times 0.15 = 0.47$ m s^{-1}

(b) Acceleration $= -\omega^2 y = -\pi^2 \times 0.10 = 0.98$ m s^{-2} and

speed $= \omega\sqrt{A^2 - y^2}$

$= \pi\sqrt{0.15^2 - 0.10^2} = \pi \times 0.12 = 0.35$ m s^{-1}

(c) $y = A \sin \omega t$

$\therefore$ $0.10 = 0.15 \sin \pi t$

$\therefore$ $\pi t = \sin^{-1} 0.6666 = 0.7297$ rad

$\therefore$ $t = \frac{0.7297}{3.142} = 0.23$ s

Test Yourself

1 A mass on a spring is performing a simple harmonic motion with a frequency of 10 Hz and amplitude 20 mm. Calculate:

(a) its displacement and (b) its velocity at a time 0.012 s after passing through the centre of oscillation.

HINT *Use $y = A \sin \omega t$ in radians.*

2 The depth of water in a certain harbour varies with the tide from 20 m to 26 m with a period of 12 hours 26 minutes (12.43 hours). Low tide on a particular day was at 9 a.m.

(a) What is the amplitude of the depth variation?
(b) At what time of that day is the depth at its middle value?
(c) At what time does the depth reach a value of 24 m?

HINT *Displacement is zero with level rising at 3.1 hour after 9 a.m. Amplitude is 3 m. You can work in hours.*

15.3 Sines of angles greater than 90°

Another look at simple harmonic motion

Sines of angles greater than 90° are important for sine curves and sine waves in this chapter and are relevant to the SHM already discussed.

For the **SHM** of the shadow explained above the downward movement from the top of the motion towards its centre (M) in *Fig. 2* is a perfect reversal of the preceding rise. So the formula $v = \omega\sqrt{A^2 - y^2}$ needs to give the same answer for the downward movement as for the upward except for a reversal of sign. As it happens the square root causes a ± answer and so this equation applies to the downward motion. But how does the formula $v = R\omega \cos \omega t$ fit the downward motion bearing in mind that angle ωt or θ is greater than 90° and no meaning has been given to the trigonometric ratios of such large angles in this book so far? A cosine has been defined as adjacent side over hypotenuse in a right-angled triangle and in such a triangle there cannot be an angle greater than 90°. Will the formula $y = A \sin \omega t$ apply when ωt exceeds 90° (or $\frac{\pi}{2}$ radian)?

There is a need to redefine the meanings of **sin**, **cos** and **tan** to deal with angles greater than 90° but in such a way as to agree with the definitions already used.

New definitions for trigonometric ratios

In *Fig. 3(a)* a line of length R makes an angle θ with the x-axis. Sin θ in that figure is the ratio of the y value to the R value and $\sin \theta = a/R$.

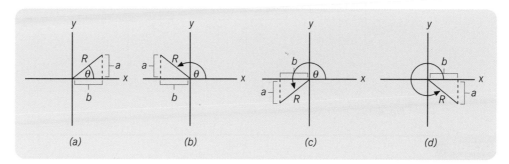

Fig. 3 *Trigonometric ratios for angles beyond 90°.*

In *Fig. 3(b)* θ is between 90° and 180° and we define sin θ as the y value divided by R as before but now have to consider signs for a and b values. a is given a + sign if it is above the x-axis and negative if below, as in a graph, while b is to be + if it is to the right of the y-axis and negative if it is to the left. R is always regarded as positive. For our sin θ in *Fig. 3b* the y value is $+a$ and R is positive so sin θ is positive and equal to $+\frac{a}{R}$.

In *Fig. 3(c)* the y value is negative so that $\sin \theta = -\frac{a}{R}$ and in *Fig. 3(d)* it is negative and so sin θ is negative.

In contrast, for θ between 90° and 180° as in *Fig. 3(b)* the x value is negative so a negative sign is placed before b and $\cos \theta = -\frac{b}{R}$ and in *Fig. 3(c)* where θ is between 180° and 270° the x value is negative so that $\cos \theta = -\frac{b}{R}$.

KEY FACT   *For any angle,* $\sin = \frac{y \, value}{R}$, $\cos = \frac{x \, value}{R}$ *and* $\tan = \frac{y \, value}{x \, value}$.

Returning to simple harmonic motion the definitions of the trigonometric ratios for angles beyond 90° agree with all that has been said. For example the equation $v = R\omega \cos \omega t$ gives a negative value when ωt is between 90° and 180° i.e. during the descent towards the middle and a negative value again for ωt between 180° and 270° when the movement is downward from the middle. Similarly the equations $A = -R\omega^2 \sin \omega t$ and $y = A \sin \omega t$ give the correct signs for a and y. Using the new definitions we get for example sin 110°= 0.9397, sin 184°= −0.0698, cos 110°= −0.3420 and cos 184°= −0.9976. Check these on your calculator.

Example

1 Given that sin 20° = 0.3420 find without using a calculator the values of:

    (a)   sin 160°   (b)   sin 200°   (c)   sin 340°

Answer

In *Fig. 4(a)* the line defining the 20° angle is chosen to be one unit long so that the length opposite the 20° angle is 0.3420 units.

(a)   In *Fig. 4b* sin 160°= + 0.3420/1 = +0.3420

(b)   In *Fig. 4c* sin 200°= −0.3420/1 = −0.3420

(c)   In *Fig. 4d* sin 340°= −0.3420/1 = −0.3420

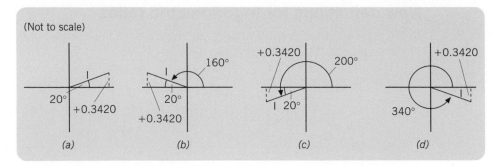

(Not to scale)

(a)   (b)   (c)   (d)

Fig. 4

1 State whether the named trigonometric ratio is + or − for the angle range given:

(a) sine between 0 and 90°   (b) sine between 90° and 180

(c) sine between 180° and 270°.   (d) cosine between 180° and 270°

(e) tangent between 90° and 180°   (f) tangent between 180° and 270°.

2 Write down the value of each of the following trigonometric ratios. (No need for a calculator.)

(a) $\sin 180°$ (b) $\cos 0$ (c) $\sin 2\pi$ (radians)   (d) $\cos \pi$ (radians)

3 Given that $\cos 60° = 0.5$ write down without using a calculator the value of:

(a) $\cos 120°$ (b) $\cos 300°$

4 Using a calculator discover the values of:

(a) $\sin 70°$ (b) $\sin 140°$ (c) $\sin 290°$

15.4 Sine curves and sine waves

Sine curves

The graph of y versus x when y is proportional to $\sin x$ is shown in *Fig. 5* and is called a **sine curve**. For a SHM $y = A \sin \omega t$ so the graph of the displacement y versus ωt or even versus t is a sine curve. A relationship that has a sine curve graph is said to be **sinusoidal**.

KEY FACT   *For SHM the displacement varies sinusoidally with time. The y versus t graph is a sine curve.*

Alternating voltages and the currents they produce are often sinusoidal and are described by the formulae $V = V_p \sin 2\pi ft$ and $I = I_p \sin 2\pi ft$. Often the alternating voltage is produced by a generator in which a rotor turns so that there is a circular motion involved and the sinusoidal nature of the voltage is not then too surprising.

For any process that repeats in a regular manner the part that repeats is called a **cycle** and the time taken for each cycle to be completed is, the **period** denoted by T. For a sine curve a rise from zero, return to zero, increase in opposite direction and return to zero is one cycle exactly. Equally the curve from a crest to trough to the next crest is one cycle. The period T is shown in *Fig. 5*.

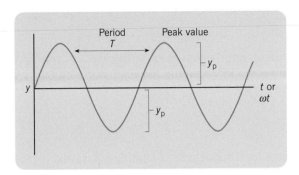

Fig. 5 *A sine curve.*

Waves in A-level physics

The waves seen on the sea are **progressive** waves because a rise or fall in water level at one place moves forward to affect other places. **Standing waves** (or **stationary waves**) are different and are met in your physics work. We will discuss the mathematics associated with progressive waves on water.

As *Fig. 6(a)* shows, at a given place where waves are present the water level rises and falls with *time*. The level is alternately above and then below the normal sea-level. The movement is a simple harmonic motion. The waves as seen in *Fig. 6* are **sinusoidal**, they are **sine waves**.

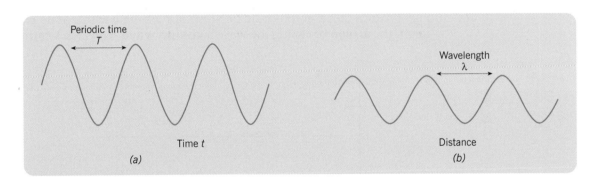

Fig. 6 *Graphs for a progressive wave.*

In *Fig. 6(b)*, which applies to a given moment in time, there is an alternate rise and fall with *distance*. The graph is identical to what you see at any one instant as you stand on the sea-shore. For a later time the graph would be displaced to the right in the same way as you see a wave on the sea move towards the shore.

KEY FACT *For a progressive wave there is an alternation, for instance of water level, with time at a given place and with distance at a given time.*

The number of cycles occurring at a point or number of waves passing any point per second equals the **frequency** of the wave.

The vertical displacement y of water level goes through a complete cycle of rise and fall f times per second, f being the frequency and this is the number of cycles passing any point per second. Each cycle takes a time T (the period, see *Fig. 6(a)*). The distance occupied by a cycle is called the wavelength and is denoted by λ (see *Fig. 6(b)*).

> **KEY FACT** *Wavelength is denoted by λ.*

The velocity or speed with which the waves move forward is denoted by c and equals $f \times \lambda$ because f cycles pass any point in 1 second and each cycle has a length λ so the length of wave passing per second is $f \times \lambda$.

> **KEY FACT** *Wave velocity $c = f\lambda$.*

At any one place the water level movement is simple harmonic (or *sinusoidal*) and we have $y = y_p \sin 2\pi ft$ where y_p is the peak value of y (the height of a wave crest) and f is the frequency. The time t is measured from any moment when $y = 0$. So the equation for the graph of *Fig. 6(a)* is $y = y_p \sin 2\pi ft$ at any one place.

For the wave shown in *Fig. 6(b)* a distance x measured from a place where $y = 0$ has been covered, in a time of $\frac{x}{c}$ or $\frac{x}{\lambda f}$, the c being the speed of travel. So the y value at this distance x is given by $y = y_p \sin \omega t$ or $y = y_p \sin(\omega \times \frac{x}{\lambda f})$ or $y = y_p \sin \frac{2\pi x}{\lambda}$

> **KEY FACT** *$y = y_p \sin \frac{2\pi x}{\lambda}$ at any one time.*

This is the equation for the graph of *Fig. 6(b)*.

It is always convenient to consider graphs of waves in which $y = 0$ when $t = 0$ and this fits our describing the waves as sine waves with $y = y_p \sin 2\pi ft$ and $y = 0$ when $t = 0$.

Fig. 7 shows that a graph could be drawn for example with y at its peak when $t = 0$. You might think then of calling the wave a **cosine wave**. We stick to **sine wave**.

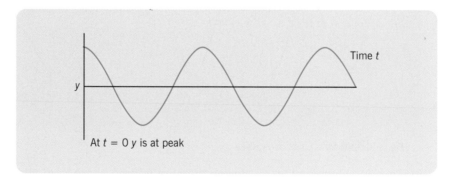

At $t = 0$ y is at peak

Fig. 7 *A wave starting with $t = 0$ when y is at a peak.*

Phase differences

In *Fig. 8(a)* there are two waves that have the same velocity and frequency (and so same period and wavelength) but each peak of wave 2 occurs later than the corresponding peak of wave 1. The waves are not **in phase**, they are not in step. One is always at a later stage in its cycle.

In *Fig. 8(b)* the waves are in exactly opposite phase, one being at a positive maximum when the other is at a negative maximum, a crest in one occurring with a trough in the other. Here the phase difference between a peak of wave 2 and the following or preceding peak in wave 2 is a half period ($T/2$). This half cycle phase difference can also be described as 180° or π radians.

> **KEY FACT** *Two waves with opposite phase have a phase difference of 180° or π radians or a half-cycle.*

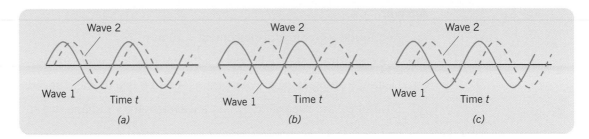

Fig. 8 *Phase differences.*

In *Fig. 8(c)* wave 2 *lags* behind wave 1 by a quarter cycle or 90° or $\frac{\pi}{2}$ radians. You could of course regard wave 2 as $\frac{3}{4}$ of a cycle in advance of wave 1 but unless there is good reason for doing this the smaller difference would be quoted.

If the equation for wave 1 in *Fig. 8(c)* is $y = y_p \sin \omega t$ then the equation for wave 2 which is $\frac{\pi}{2}$ behind wave 1 is $y = y_p \sin(\omega t - \frac{\pi}{2})$. In this equation y_p is chosen to be the same as for wave 1 only because the amplitudes have been made the same in *Fig. 6(c)*.

Why the minus sign? Well, at a time t the y value in wave 2 will be the same as that for wave 1 at an earlier time, in fact earlier by an angle $\frac{\pi}{2}$ or time $\frac{T}{4}$. So y for wave 2 is the same as y in wave 1 at an angle of $\omega t - \frac{\pi}{2}$. Thus the equations are

KEY FACT $y = y_p \sin \omega t$ *for wave 1 and* $y = y_p \sin(\omega t - \frac{\pi}{2})$ *for wave 2 where* $\frac{\pi}{2}$ *is the lag of wave 2 on wave 1.*

Example

(a) In *Fig. 9* graph A shows the displacement plotted against distance travelled for a progressive wave on a string. Graph B applies to the same wave at a time 0.20 s later.

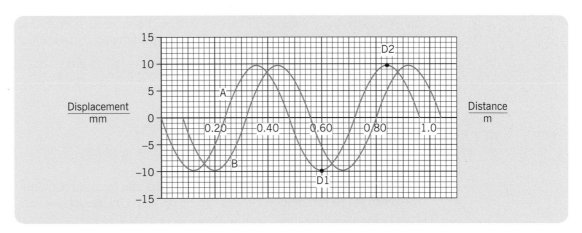

Fig. 9

Use the graphs to determine,

(a) the amplitude of the wave

(b) its wavelength

(c) its speed of travel

(d) the phase difference between points D1 and D2 on graph A.

Answer

(a)   Amplitude = greatest displacement = 10 mm

(b)   Wavelength = distance between adjacent peaks in wave = 0.86 − 0.36 = 0.50 m

(c)   Distance along axis between graphs A and B is 0.08 m so speed = $\frac{0.08}{0.20}$ = 0.40 m s^{-1}

(d)   D1 and D2 are separated by half a wavelength, π radians or 180°.

Test Yourself

Exercise 15.4.1

With reference to *Fig. 10*, what is the phase relationship between wave A and wave B

    (a) in *Fig. 10(a)*   (b) in *Fig. 10(b)*   (c) in *Fig. 10(c)*?

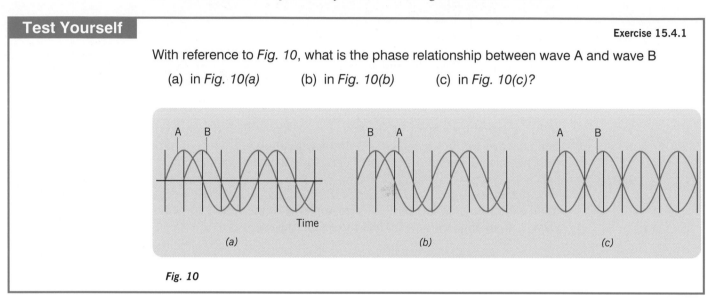

Fig. 10

Complete equation for a progressive wave

The complete equation for a wave tells us about the variation in *y* with time and with distance. At any given place receiving a sine wave $y = y_p \sin \omega t$. Measuring distance *x* forward from a chosen place the *y* value at distance *x* will be the same as occurred earlier at *x* = 0, earlier by a time of $\frac{x}{c}$ which is $\frac{x}{\lambda f}$. Hence *y* at position *x* is given by an equation not containing simply ωt

but by $y = y_p \sin \omega \left(t - \dfrac{x}{\lambda f} \right) = y_p \sin\left(\omega t - \dfrac{\omega x}{\lambda f} \right) = y_p \sin\left(\omega t - \dfrac{2\pi x}{\lambda} \right).$

KEY FACT   *For any point on a sine wave* $y = y_p \sin\left(\omega t - \dfrac{2\pi x}{\lambda} \right)$

As seen above there are many expressions that can be used for *y* but this last one is probably the best one to remember. Others can be deduced by use of $\omega = 2\pi f$, $\lambda f = c$ and $T = \frac{1}{f}$.

Example

The equation $y = 0.030 \sin 2\pi(2.0t + x)$ describes a progressive wave moving along a string. *y* is the displacement at time *t* of a point on the string at a distance *x* from the origin. *y* and *x* are measured in metres, *t* in seconds. Deduce values for (a) the amplitude, (b) the frequency (c) the wavelength (d) speed of travel of the wave.

Answer

(a)   The amplitude is the maximum value of *y* that occurs when the sine reaches its maximum value of 1 (angle = 90°). So amplitude = 0.030 m.

(b) The given equation has the form of $y = y_p \sin(\omega t - \frac{2\pi x}{\lambda})$ in which $\omega = 2\pi f$ but the 2π present in each of the terms in the brackets has been put outside the brackets to give $y = y_p \sin 2\pi(ft - x/\lambda)$
Comparing this equation with the one given shows that the frequency f is 2.0, i.e. 2.0 Hz.

(c) Again comparing these two equations shows that, neglecting the − sign, the x/λ in one equation equals the x in the other. So λ must equal 1, i.e. 1 metre.

(d) The wave's speed is given by $c = f\lambda$ so $c = 2.0 \times 1 = 2.0 \text{ m s}^{-1}$.

Test Yourself

Exercise 15.4.2

1 A wave in a rope is described by the equation $y = 0.002 \sin \dfrac{2\pi}{3}(9t - x)$ where y is the displacement in metres, t is time and x is distance along the wave. All quantities are measured in the appropriate SI units. Obtain values for (a) the amplitude, (b) the frequency, (c) the wavelength, (d) the velocity of the wave.

HINTS AND TIPS

(a) (i) In text y_p was peak y or amplitude. Also see the worked example on pages 216–217

(iii) Compare with $y = A \sin\left(\omega t - \dfrac{2\pi}{\lambda} x\right)$.

15.5 Phasors

Phasor diagrams for alternating voltages

An alternating current is a to-and-fro movement of electric charge. It is an oscillation and unless otherwise stated it is simple harmonic. The phases of alternating currents and voltages causing them are important. A quantity that has phase as well as size is called a **phasor** whereas a vector has direction and size.

Just as a clock may indicate the time of day by having a rotating hand so the phase of an alternating voltage may be indicated as in *Fig. 11* by a 'hand' rotating.

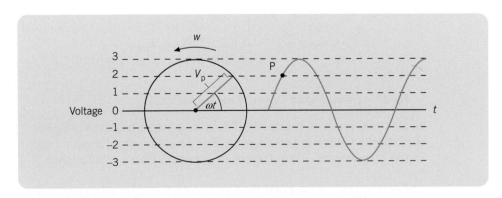

Fig. 11 A phasor diagram (or phase diagram).

This 'hand' is usually considered to rotate anticlockwise. The angle of rotation is the phase angle θ or ωt. An extra feature of this voltage clock is that the length of the hand is made to represent the peak value V_p of the voltage. What has been referred to as the 'hand of the clock' is a **phasor** because it has size V_p and phase (angle θ).

KEY FACT	*A phasor has phase as well as size.*

Since $V = V_p \sin \omega t$ the height of the tip of the hand in *Fig. 11* represents V. A number of horizontal lines can be drawn as in the figure to indicate voltage values so that V could be read from the clock at any time. Also shown in *Fig. 11* is the sine curve relating V to t. On this curve the point P has been reached corresponding to the reading of the voltage clock.

A phasor diagram for combining voltages

In A level physics, calculations are often met that involve an inductor or capacitor in a circuit, perhaps both, usually with resistance present too. Just such a circuit is represented by *Fig. 12*.

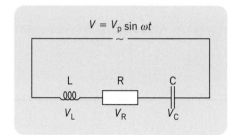

Fig. 12 *An LCR series circuit.*

The supply voltage is sinusoidal. L, R and C are quantities that are characteristic of the components that they describe, namely the inductance of a coil, resistance and capacitance respectively. The voltages at any instant across L, R and C do not all have the same phase. *Fig. 13* shows a phasor diagram that represents the voltages across L, R and C.

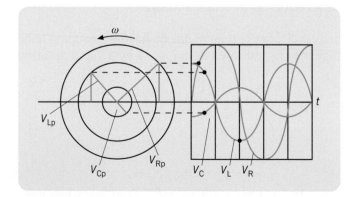

Fig. 13 *A phasor diagram for an L, R and C series circuit.*

As all three phasors rotate together (like a propeller) the voltages V_L, V_R and V_C are given by the heights shown in the diagram (*Fig. 13* again) and the sine curves for these voltages drawn in the diagram are just sufficient to show the phase relationships.

These phasors can be combined and so replaced by a single phasor just as vectors are combined. The resultant of the L and C phasors is obtained by subtraction and their resultant is combined with the R phasor at right angles to it by using the Pythagoras formula. The resulting voltage must be equal to the supply voltage and this fact enables formulae to be obtained for calculations involving alternating currents and voltages.

Exam Questions

Exam type questions to test understanding of Chapter 15

Exercise 15.4.3

1 An aircraft flies with its wings tilted as shown in *Fig. 14* in order to fly in a horizontal circle of radius r. The aircraft has mass 4.00×10^4 kg and has a constant speed of 250 m s^{-1}.

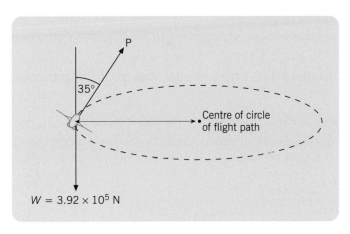

Fig. 14

With the aircraft flying in this way, two forces acting on the aircraft in the vertical plane are the force P acting at an angle of 35° to the vertical and the weight W.

(a) State the vertical component of P for horizontal flight.
(b) Calculate P.
(c) Calculate the horizontal component of P.
(d) Use Newton's second law to determine the acceleration of the aircraft towards the centre of the circle.
(e) Calculate the radius r of the path of the aircraft's flight.

(OCR 2000)

> **HINT**
>
> *For (a) and (c) see Chapter 11. Newton's second law equation is acceleration $= \frac{force}{mass}$. For (e) v^2/R = inward acceleration.*

2 A motorist notices that when driving along a level road at 95 km h^{-1} the steering wheel vibrates with an amplitude of 6.0 mm. If she speeds up or slows down, the amplitude of the vibrations becomes smaller.

(a) Explain why this is an example of resonance.
(b) Calculate the maximum acceleration of the steering wheel given that its frequency of vibration is 2.4 Hz.

(Edexcel 2001)

> **HINT**
>
> $a = -\omega^2 y$, $\omega = 2\pi f$. Max a at $y = A$.

3 A particle is oscillating with simple harmonic motion described by the equation

$$s = 5 \sin (20\pi t)$$

How long does it take the particle to travel from its position of maximum displacement to its mean position?

A $\frac{1}{40}$ s   **B** $\frac{1}{20}$ s   **C** $\frac{1}{10}$ s   **D** $\frac{1}{5}$ s

(AQA 2000)

> **HINT**   *Assume 20π is ω in rad s^{-1}.*

4 A progressive transverse wave travelling in the negative x direction along a rubber rope is represented by

$$y = 6 \times 10^{-3} \sin(8\pi t + 4\pi x) \text{ where all quantities are in the appropriate SI units.}$$

(a) Find:

    (i) the amplitude of the wave   (ii) the wavelength.

(b) Draw a displacement-distance graph for the wave along the rope at time $t = 0$ using a copy of the axes below (*Fig. 15*).

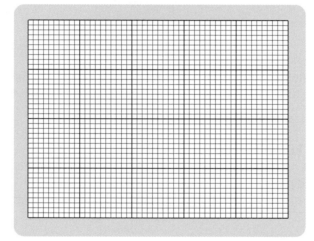

Fig. 15

(c) How far apart are two successive points on the wave which have a phase difference of $\frac{\pi}{2}$ radians?

(WJEC 1999)

> **HINT**
> (a) (i) *Assume y is displacement (in metres).* (ii) $\frac{2\pi}{\lambda} = 4\pi$.
> (b) *For $t = 0$, $y = 6 \times 10^{-3} \sin 4\pi x$ so that $y = 0$ when $x = 0$. One wave is completed when $x = \lambda$.*
> (c) $\frac{\pi}{2}$ *means a quarter of a cycle*

Answers to Test Yourself Questions

Exercise 15.1.1
1 314 rad s^{-1}
2 (a) 14 m s^{-1} (b) 51 km h^{-1}

Exercise 15.1.2
1 (a) 10 m s^{-1} (b) 40 N

Exercise 15.2.1
1 (a) 14 mm (b) 916 mm s^{-1} or 0.92 m s^{-1}
2 (a) 3.0 m (b) 6 min past 12   (c) 14 min to 1 pm

Exercise 15.3.1
1 (a) +   (b) +   (c) −   (d) −   (e) −   (f) +

2 (a) 0   (b) 1   (c) 0   (d) −1
3 (a) −0.5   (b) 0.5
4 (a) 0.9397   (b) 0.6428   (c) −0.9397

Exercise 15.4.1
1 (a) B lags behind A by 90° or $\frac{1}{4}$ cycle or $\pi/2$ radian.
   (b) B is in advance or A lags on B by $\frac{1}{4}$ cycle.
   (c) 180° or π radian, out of phase.

Exercise 15.4.2
1 (a) (i) 0.002   (ii) 3 Hz   (iii) 3 m   (iv) 9 m s^{-1}

Appendices

Appendix I

Combinations of Experimental Errors

Error in a product

> **KEY FACT** *For a product we calculate the* percentage *error by* adding percentage *possible errors.*

Let's first look at an example:

> A rectangle has a length of 100 cm ± 1 cm and its width is measured as 100 ± 1 cm. So its area can be as small as $99 \times 99 = 9801$ cm^2 or as big as $101 \times 101 = 10\,200$ cm^2. This means that the area is, to a close approximation $10\,000 \pm 200$ cm^2. As a percentage of 10 000 the 200 is $200/10\,000 \times 100\% = 2\%$. Therefore the area is 10 000 cm^2 ± 2%. This compares with ±1% in each of the two multiplying quantities.

In general if two quantities x and y are measured with possible errors of δx and δy then the calculated product is $(x \pm \delta x)(y \pm \delta y)$ and the possible error in the product is

$$(x \pm \delta x)(y \pm \delta y) - xy$$

Multiplying out the brackets as explained in Chapter 3 gives

$$\text{possible error in product} = \delta p = xy \pm x\,\delta y \pm y\,\delta x \pm \delta x\,\delta y - xy$$

The xy terms cancel and if we divide the equation by xy and multiply by 100 we get

$$\frac{\delta p}{xy} \times 100 \quad \text{or} \quad \frac{\delta p}{p} \times 100 = \pm\left(\frac{\delta y}{y} \times 100\right) \pm \left(\frac{\delta x}{x} \times 100\right) \pm \left(\frac{\delta x\,\delta y}{xy} \times 100\right)$$

We can assume the errors δx and δy to be small compared with x and y and this means that the term containing $\delta x\,\delta y$ is much smaller than the other terms and can be neglected. (For example, if δx and δy were each a hundredth of x and y then the terms on the right of the last equation would be ±1 ±1 ± 0.01 and the 0.01 can be neglected.)

The equation becomes $\dfrac{\delta p}{p} \times 100 = \pm \dfrac{\delta y}{y} \times 100 \pm \dfrac{\delta x}{x} \times 100$ which tells us that the percentage possible error in a product equals the sum or difference (depending on the signs of the errors) of the percentage possible errors of its factors. Because we want the maximum error that could possibly occur we must think of two + signs or two − signs and write

$$\frac{\delta p}{p} \times 100 = \pm\left(\frac{\delta y}{y} \times 100 + \frac{\delta x}{y} \times 100\right)$$

or

> % possible error in product = ±*sum* of percentage possible errors of its factors

For one quantity dividing the other, $\frac{x}{y}$, a similar rule can be deduced. You might expect it to show that the percentage errors would subtract but because of the presence of ± signs the percentage errors do add. So the above rule applies. You add the percentage errors in x and y.

In contrast for the *sum* of two quantities x and y the error in the sum is

$(x \pm \delta x) + (y \pm \delta x) - (x + y)$ which works out to $\pm\delta x \pm \delta y$.

As before we must think of two + or two − signs and so the (maximum) possible error in the sum equals ± the sum of the absolute errors δx and δy.

For a sum we calculate the **absolute** error by adding absolute errors

For a difference $x - y$ we get the same result, *add absolute errors*.

Appendix II

Proof of Pythagoras' equation

This important equation discovered by Pythagoras relates the length a of the hypotenuse of a right-angled triangle to the lengths b and c of the triangle's other two sides.

$$a^2 = b^2 + c^2$$

In the figure a right-angled triangle is drawn and for the purpose of deriving the equation a line AD has been added which is perpendicular to BC.

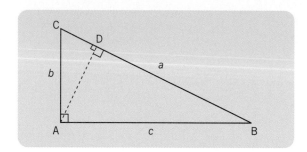

Diagram for proof of Pythagoras' equation.

The small triangle ACD containing the length CD is similar (see Chapter 10) to the large triangle ABC and so the $\frac{\text{adjacent side to angle C}}{\text{hypotenuse}}$ is the same for both triangles.

For the triangle ACD this ratio is $\frac{CD}{b}$ while for large triangle ABC it is $b\,/\,a$.
Therefore $\frac{CD}{b} = \frac{b}{a}$ or $CD = \frac{b^2}{a}$.
For the triangle ABD the ratio $\frac{\text{adjacent side to angle B}}{\text{hypotenuse}} = \frac{BD}{c}$ while for large triangle ABC it is $\frac{c}{a}$.
Therefore $\frac{BD}{c} = \frac{c}{a}$ or $BD = \frac{c^2}{a}$.
Now $a = CD + BD$ so that $a = \frac{b^2}{a} + \frac{c^2}{a}$ and multiplying both sides by a gives $a^2 = b^2 + c^2$ as required.

Appendix III

Tables of symbols, abbreviations and units

Table I *Abbreviations and symbols*

Abbreviations for SI units		Mathematical symbols	
kilogram (mass)	kg	equals	$=$
metre (length)	m	identical to	$\equiv$
second (time)	s	proportional to	$\propto$
hertz (frequency)	Hz	square root	$\sqrt{}$
newton (force)	N	of the order of ('something like')	$\sim$
joule (work or energy)	J	divided by	$/$
watt (power)	W	therefore	$\therefore$
kelvin (temperature)	K	is greater than	$>$
volt (potential difference)	V	is less than	$<$
ampere (current)	A	is much greater than	$\gg$
ohm (resistance)	Ω	infinity	∞
farad (capacitance)	F	a small increase in	δ or Δ
henry (inductance)	H	the sum of all values of x	Σx
tesla (magnetic flux density)	T	pi (see page 145)	π
weber (magnetic flux)	Wb	exponential function	e
coulomb (charge)	C		
becquerel (activity)	Bq		

Table II *Multiples and submultiples for units*
The following prefixes are commonly used and their values are given below

Symbol	Prefix	Value	In words
p	pico	10^{-12}	a million-millionth
n	nano	10^{-9}	a thousand-millionth
μ	micro	10^{-6}	a millionth
m	milli	10^{-3}	a thousandth
c	centi	10^{-2}	a hundredth
d	deci	10^{-1}	a tenth
k	kilo	10^{3}	a thousand times
M	mega	10^{6}	a million times
G	giga	10^{9}	a thousand million times

Answers to exam-type questions

Exercise 1.2.8
1 **B**
2 **A**
3 1 728 000 joule
4 8
5 (a) 24 watt   (b) 6 ohm

Exercise 2.5.2
1 112 000 joules or 112 kJ or 1.12×10^5 J
2 1.5 μF
3 $\frac{5}{9}$
4 (a) $\frac{49}{100}$ (b) $\frac{1}{343}$

Exercise 3.3.1
1 5.6 ohm
2 1500 watt
3 1.2 m

Exercise 4.4.2
1 14 N m^{-1}
2 **A**
3 **B**
4 33%
5 (i) 2.4×10^{-10} farad   (ii) 1.2×10^{-7} C
   (iii) 3.0×10^{-5} J

Exercise 5.2.4
1 **A**. Best answer without taking mean is
   $1.04 \pm 0.01 \times 2$ meaning between 1.04 and 1.06.
   1.04 would imply 1.04 ± 0.05
2 kg m^{-1}
3 $x = \frac{1}{2}, y = -\frac{1}{2}$
4 All terms have same units. Equations can be
   incorrect due to numbers involved.

Exercise 6.5.2
1 0.57 s
2 915 K or 9.1×10^2 K or 0.91×10^3 K
3 3.3×10^2 m s^{-1} or 0.33 km s^{-1}

Exercise 7.4.2
1 (a) ±1.1%   (b) ±2.3%   (c) ±5%
2 2000 ohm ± 20 ohm
3 (a) 1.12 g cm^{-3} (b) ±0.68%

Exercise 8.3.4
1 **C**
2 (a) 1.53×10^{-17} J   (c) 0.12 nm
3 (a) $\dfrac{T_2^2}{T_1^2}$ or $\left(\dfrac{T_2}{T_1}\right)^2$ (b) 1.9 s

4 **A**

Exercise 9.4.3
1 **D**. (The line in fact cuts the temperature axis at

−273°. Answer **C** would be correct if $\theta + 273$ were
used. Not **B** because line does not pass through
(0, 0). Can't be **A** because this has no units.)
2 **A**
3 (a) Plotting T_{on} on y-axis versus $\dfrac{1}{v^2}$ (or V^{-2} on the
   x-axis is suitable. Label axes as T_{on}/ms and
   V^{-2}/V^{-2}. Expect straight line that passes
   through (0, 0).
   Alternatively can plot $\log_{10}(T_{on}/\text{ms})$ versus
   $\log_{10}(V/V)$ and expect straight line graph with
   gradient of −2 because $\log T_{on} = \log\left(\dfrac{k\eta d^2}{v^2}\right)$

   $= \log(k\eta d^2) - \log(V^{-2}) = \log(k\eta d^2) - 2 \log V$.
   (b) Graph gives straight line as expected through
   (0, 0) or with gradient −2 as stated in part (a)
   so that proportionality is confirmed.
   (c) From first graph suggested the slope is $k\eta d^2$
   and is approximately 20 ms V^2 and
   $k = 7.7 \times 10^9$ V^2 Pa^{-1} m^{-1}.
   If the log versus log graph is used then the
   intercept is 1.3 (remembering that $\log_{10}$
   (T_{on}/ms) was plotted) so that $\log_{10}(k\eta d^2)$ is 1.3
   with $k\eta d^2$ measured in ms. The antilog is
   obtained using 10^x and is 20 ms. Converting
   this to s and substituting other values given
   gives k.
   (d) Plot T_{on} versus d^2 and expect straight line
   through (0, 0) to show proportionality or
   plot $\log T_{on}$ versus $\log d$ and expect gradient
   equal to 2.
   V would have to be kept constant. k and η
   would also be constant.

Exercise 10.6.2
1 (a) 30 km s^{-1} (b) 2.0×10^{-7} rad s^{-1}
2 4.5×10^{-3} rad
3 4.7×10^{-2} W m^{-2}
4 (a) 6 protons, 8 neutrons   (b) 8.2×10^{-44} m^3
   (c) 2.8×10^{17} kg m^{-3} (f) 3.8×10^{-12} s^{-1}

Exercise 11.3.3
1 **B**
2 **D**
3 (a) 6.9 N   (b) 6.9 N
4 **A**

Exercise 12.3.3
1 5.3×10^3 kg m^{-3}
2 15 m s^{-1}, 45 m
3 **C**
4 (a) (ii) 15.5 m s^{-1}.
   (b) 202 m
   (c) 77 s

Exercise 13.3.3

1 $x = 27\ \Omega$, $y = 33\ \Omega$

2 1.0 s (Answer of 2.0 s is time for height to be 10 m after reaching maximum height and then falling.)

3 100 and 150 Ω

Exercise 14.2.5

1 (a) 322 dB   (b) -40 (c) 4

2 (b) (i) 1.443

3 (a) 2.6×10^{-2} year^{-1} or 8.1×10^{-10} s^{-1}

   (b) 3.3×10^{-6} mole

   (c) 1.6 kBq

4 (a) $\gamma = \dfrac{\log P_2 - \log P_1}{\log V_1 - \log V_2}$ (b) 1.4

Exercise 15.4.3

1 (a) $P \cos 35$ (b) 4.8×10^5 N

   (c) 2.7×10^5 N   (d) 6.9 m s^{-2}

   (e) 9.1×10^3 m

2 1.4 m s^{-2} or 1.4×10^3 mm s^{-2}

3 A

4 (a) (i) 6.0×10^{-3} m   (ii) 0.50 m

   (b) *See Fig. A5*

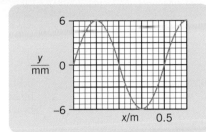

Fig. A5

   (c) 0.125 m

Index

3-dimensional shapes 146–9

A

abbreviations, units 62, 64, 225
abscissa 109
absolute errors 85, 87–9, 223
acceleration 63, 68
  circular motion 207
  due to gravity 54
  uniform 128
accuracy *see* errors; significant figures
activity 131
acute angles 137
addition 37
  directed numbers 39
  errors 88
  exponents 46
  fractions 25–6
  powers of ten 50
algebra 94–108
  see also equations
  proportionality 99–107
  ratio 103
alternating current 127, 179–80,
    190–1
  phasors 217–19
amplitude 208
angles 62, 136–40
  radians 149–50
  triangles 137–40
  trigonometric ratios 153–8, 210–12
angular frequency 209
angular velocity 150, 206
answer memory 10
antilogs 197–8
arcs 145, 149–50
area 64, 68
  circles 145
  rectangles 64, 144
  spherical surfaces 147
  triangles 143
  under graph 128–30, 174–5
averages 172–81
  best-fit line 117
axes 109, 111

B

base units 62–3, 71–2
bases
  logarithms 194, 195
  powers 44, 45, 53

best-fit line 116–17
bisectors 141
brackets 34–7, 68

C

calculations
  accuracy 90–2
  units 60, 65–6
calculators
  angles 149
  brackets 34
  decimal fractions 28
  division 6, 8–10
  errors 84
  EXP key 52
  exponential function 55
  logarithms 195–6, 198
  mean 172
  memory 10
  modes 3, 57
  multiplication 2–3
  negative numbers 41, 52
  pi (α) 145
  powers 51–3
  reciprocals 25
  roots 75–6, 78, 80
  scientific mode 46, 57, 86
  second function 51
  SHIFT key 52
  squaring 45
  trigonometric ratios 155, 156
cancelling
  equations 98
  fractions 18–20
  units 61
capital letters, units 63
checking
  calculations 4–5, 53, 79
  formulae 69–71
  units 69
chords 144
circles 144–6
circular motion 76, 206–8
circumference 145
common factors 36
common logarithms 194–5
components, vectors 158–9,
    162–7
compression 68
congruent triangles 141
constant of proportionality 101
continuous variables 174–7

conversions
  fractions/decimals 29
  units 65–7
coordinates 110
cosine (cos) 153–5, 157–8,
    210–12
cosine wave 214
cube root 75
cubed numbers 44
cubic equation 182
cuboids 65, 146–7
current 100, 109
  alternating 218
  RMS 179–80, 190–1
curves, graphs 122–4
cycle 212
cylinders 146–7, 148–9

D

data 90
decay constant 55
decimal fractions 28–30, 48
decimal numbers 3–4
decimal point 3, 28, 45
degrees 136
denominator 15
density 68, 76
dependent variable 111
derived units 63
diagrams
  angles 137
  phasors 217–18
  vectors 158–9
diameter 144
difference 37
diffraction gratings 154
dimensions 67–72, 146
direct proportionality 104
directed numbers 37–41
discs 145–6
displacements 158
distance, velocity/time graph 175
division 6–12
  by ten 7
  directed numbers 40
  errors 88
  fractions 22–3
  powers 48–9
  using logs 199
drawing
  see also diagrams
  graphs 109–12, 116–17

E

e *see* exponential function
equals sign 1
equations 95–9
  proportionality 101–2
  quadratic 188–92
  rearranging 95–8
  simultaneous 183–8
  solving 99, 186–92
  square roots 77
  straight line graphs 119–20, 125–7
  types 182
equilateral triangles 140–1
equilibrium 168–9
equipment, errors 83
error distribution curve 173, 177–8
errors 83–93
  combining 87–9, 222–3
  significant figures 86–7, 90–1
  types 83–5
estimation 4–5, 53, 79
Euler's number *see* exponential function
EXP key 52
experimental errors 83–5, 222–3
experimental results
  error distribution curve 172–3, 177–8
  graphs 110, 116–17, 200–3
  powers of ten 51
  standard deviation 179
  units 61
exponential change 201
exponential function (e) 54–6, 131
  logarithms 195
exponential relationships 55–6
exponents 44
  addition 46
  fractional 78
  pure numbers 69
  subtraction 48
  variable 55, 200–1
expressions 8
  decimals 30
  fractions 17–21
exterior angles 137–8
extrapolation 125

F

factors 1, 36
flow rate 54, 78
forces 12, 60, 63
  circular motion 207
  combining 159–62
  current carrying wires 106, 117–18
  dimensions 68, 71
  parallelogram 159–60
  resolving 162–7
  triangle 168–9
formulae 95, 126
  circular motion 206
  dimensions 69, 70–2
  quadratic equations 188–9

  simple harmonic motion 208–9
  units 54
  waves 216
fractions 15–33
  addition 25–6
  cancelling 18–20
  decimal 28–30
  division 22–3
  exponents 78
  multiplication 21–3
  names 15–16
  proper/improper 15
  reciprocals 24–5
  subtraction 26
frequency 150
  angular 209
  circular motion 206
  wave 213
functions 184

G

g (acceleration due to gravity) 54
gradients 113–17
  curves 122–3
  exponential function 131
  negative 121–2
graphs 109–35
  area under line 128–30, 174–5
  axes 109, 111
  curves 122–4
  drawing 109–12, 116–17
  equations 119–20, 125–7
  exponential function 131
  gradient 113–17, 121–3
  graph paper 109
  logarithmic scales 196–7, 200–3
  non-linear 132–3
  predictions 124–5
  proportionality 100–1, 112–15, 119
  quadratic equations 192
  roots 132–3
  scales 109, 111–12
  simultaneous equations 186–8
  straight-line 100–1, 112–22
  units 61–2, 111–12
greater than 84
Greek symbols 94, 137

H

half-life 55–6
hertz (Hz) 150
hypotenuse 141, 153
Hz *see* hertz

I

idealised point 117
identities 95
identity sign 68
impedance 190–1

improper fraction 15
independent variable 111
index *see* exponents
indices *see* exponents
infinity 199
intercepts 115
interior angles 137–8
interpolation 125
intersection, graphs 187
inverses
  antilogs 197–8
  proportionality 104
  trig ratios 155–6
isosceles triangles 140

J

joule 63

K

kilogram 62

L

labels, axes 111
lag 215
length 62
less than 84
linear *see* straight-line
ln *see* natural logarithms
logarithmic change 201
logarithms 194–205
  antilogs 197–8
  base 10 194–5
  graphs 196–7, 200–3
  natural 195–6
  rules 198–9

M

magnitude 159, 179
mass 62
mean 172–7
measurements 60–1
  see also units
  errors 83–5
memory, calculators 10
metre 62
modes, calculators 3, 57, 86
mole 45
moments 13, 118
momentum 68–9
multiples, units 45, 64, 225
multiplication 1–6
  brackets 34–7
  by ten 4
  directed numbers 39–40
  errors 87
  fractions 21–3
  powers 46–8
  using logs 199

N

N *see* newton
natural logarithms 195–6
negative numbers 37–41
    powers 48, 52–3
    roots 80–1
newton (N) 60, 63
normal distribution 177–8
normal lines 137
numerator 15

O

obtuse angle 137
order
    division 7
    multiplication 1
ordinary logarithms 194–5
ordinate 109
origin 110
oscillations *see* simple harmonic motion;
        waves

P

parallelogram rule 159–60
parallelograms 144
p.d. *see* potential difference
peak values 180, 213
pendulum 125–6, 209–10
per 11
percentage 30–1
percentage errors 85, 87–9, 222
period 150, 206, 212–13
perpendicular 137
perpendicular bisector 141
phase differences 214–15
phasors 217–19
pi (π) 145
positive numbers 38, 80–1
possible errors 84
potential difference (p.d.) 24, 100
power, electrical 13
powers 44–59
    arithmetic 46–51
    bases 44, 45, 53
    negative 48, 52–3
    roots 78
    standard form 56–7
    unknown 202
    variables 53–4, 55
    zero 49
precision 83
predictions, graphs 124–5
prefixes, units 45, 64, 225
probable error 84
products 1, 36–7
progressive waves 213–17
proper fraction 15
proportionality 99–107
    direct 104
    equations 101–2

graphs 100–1, 112–15, 119
    inverse 104
    ratio 103
pure numbers 61, 68
Pythagoras' theorem 141–2, 161, 224

Q

quadratic equations 188–92
quadrilaterals 144
quantity 61

R

radians (rad) 149–50
radioactivity 55–6, 131, 147–8
radius 144
random errors 83
range 173
rate of change 115
ratios 103
    trigonometric 153–8, 210–12
reaction 68
rearranging equations 95–8
reciprocals 24–5
    logarithms 199
    powers 48
rectangles 64, 144
recurring decimals 28
reflex angle 137
refraction 156
relationships, sine/cosine 157–8
relative errors 85
resistance 15, 95, 101–2, 116
resistivity 69
resolving, forces 162–7
resultant, vectors 158–62
results tables 51, 61
revolution 136, 149
right-angle 136
right-angled triangles 141–2
RMS *see* root mean square
root mean square (RMS) 177–80
roots 74–82
    calculations 77, 79–80
    graphs 132–3
    positive/negative numbers 80–1
    powers 78
rough checks 4–5, 53, 79
rounding 91
rules
    directed numbers 39–40
    logarithms 198–9
    powers 47, 48
    rearranging equations 96, 98

S

scalar quantity 158
scales
    graphs 109, 111–12
    linear 197

logarithmic 196–7, 200–3
scientific mode 46, 57, 86
second 62
second function 51
sector 144
s.f. *see* significant figures
sharing *see* division
SHIFT key 52
SHM *see* simple harmonic motion
SI (Système Internationale) units 62–5
significant figures (s.f.) 28, 86–7, 90–1
signs
    see also symbols
    directed numbers 38–9
similar triangles 141
simple harmonic motion (SHM) 208–12
simple pendulum 125–6, 209–10
simplifying
    decimal fractions 30
    fractions 17–21
simultaneous equations 183–8
sine curve 212–13
sine (sin) 153–8, 210–12
sinusoidal relationship 212
slope
    curves 122–3
    exponential function 131
    negative 121–2
    straight-line graph 113–17
solving equations *see* equations
speed 11
    see also velocity
    angular 150
    rotation 150, 206
spheres 146–7
springs 77, 105, 209
square root 74–7
squared numbers 41, 44
squares 144
standard deviation 178–80
standard form 56–7, 86
standing (stationary) waves 213
straight-line equations 119–20, 125–7
    simultaneous 183–7
straight-line graphs 101, 112–22
    gradient 113–17
submultiples, units 64, 225
subtended 150
subtraction 37
    directed numbers 39
    errors 88, 92
    exponents 48
    fractions 26
    powers of ten 50
sum 37
symbols 94, 225
Système Internationale (SI) units 62–5

T

tangent, curves 122–4
tangent (tan) 153–5, 210–12
tension 68

therefore 66, 96
time 62
transposition 95
triangle of forces 168–9
triangles 137–43
   angles 137–40
   area 128, 143
   Pythagoras' theorem 141–2
   types 140–1
trigonometric ratios 153–8
   angles >90° 210–12
   inverse 155–6
   small angles 154

U

uniform acceleration 128
units 60–73
   calculations 1, 60, 65–6
   dimensions 67–72
   formulae 54

graphs 61–2, 111–12
powers 44
prefixes 45
SI 62–5

V

variables 99
   continuous 174–7
   exponents 55, 200–1
   graphs 111
   powers 53–4, 55
vectors 158–69
   combining 158–62
   directed numbers 37
   equilibrium 168–9
   resolving 162–7
velocity 11, 63
   see also speed
   angular 150, 206
   mean 174–6

RMS 179
wave 214
vertex 140
voltage *see* potential difference
voltages, alternating
   combining 218–19
   RMS 179–80, 190–1
volume 64–5, 68, 147

W

wavelength 213–14
waves 213–17
weight 6

Z

zero
   exponent 49
   logarithms 199

Acknowledgements

The author and publishers would like to thank the following awarding bodies for permission to reproduce questions from past examination papers:

Assessment and Qualifications Alliance (AQA)
Edexcel
Oxford Cambridge and RSA Examinations (OCR)
Welsh Joint Education Committee (WJEC)

Thanks are also due to Martyn Chillmaid for providing the photograph on p. 2.

**YALE COLLEGE
LEARNING RESOURCE CENTRE**